Lecture Notes in Civil Engineering

Volume 59

Lecture Notes in Civil Engineering (LNCE) publishes the latest developments in Civil Engineering - quickly, informally and in top quality. Though original research reported in proceedings and post-proceedings represents the core of LNCE, edited volumes of exceptionally high quality and interest may also be considered for publication. Volumes published in LNCE embrace all aspects and subfields of, as well as new challenges in, Civil Engineering. Topics in the series include:

- Construction and Structural Mechanics
- Building Materials
- Concrete, Steel and Timber Structures
- Geotechnical Engineering
- Earthquake Engineering
- Coastal Engineering
- Ocean and Offshore Engineering; Ships and Floating Structures
- Hydraulics, Hydrology and Water Resources Engineering
- Environmental Engineering and Sustainability
- Structural Health and Monitoring
- Surveying and Geographical Information Systems
- Indoor Environments
- Transportation and Traffic
- Risk Analysis
- Safety and Security

To submit a proposal or request further information, please contact the appropriate Springer Editor:

- Mr. Pierpaolo Riva at pierpaolo.riva@springer.com (Europe and Americas);
- Ms. Swati Meherishi at swati.meherishi@springer.com (Asia - except China - and Australia/NZ);
- Ms. Li Shen at li.shen@springer.com (China).

Indexed by Scopus

More information about this series at http://www.springer.com/series/15087

Mokhtar Awang · Meor Razali Meor M Fared
Editors

ICACE 2019

Selected Articles from the International Conference on Architecture and Civil Engineering

Springer

Editors
Mokhtar Awang
Department of Mechanical Engineering
Universiti Teknologi Petronas
Seri Iskandar, Perak, Malaysia

Meor Razali Meor M Fared
Universiti Putra Malaysia
Serdang, Malaysia

ISSN 2366-2557 ISSN 2366-2565 (electronic)
Lecture Notes in Civil Engineering
ISBN 978-981-15-1192-9 ISBN 978-981-15-1193-6 (eBook)
https://doi.org/10.1007/978-981-15-1193-6

This Springer imprint is published by the registered company Springer Nature Singapore Pte Ltd.
The registered company address is: 152 Beach Road, #21-01/04 Gateway East, Singapore 189721, Singapore

Preface

This volume presents a compilation of research works covering the fields of manufacturing and mechanical engineering. All manuscripts in this volume were presented during the 3rd International Conference on Architecture and Civil (ICACE 2019) which was held at Seri Pacific Hotel, Kuala Lumpur, Malaysia on 26–27 June 2019. The editor(s) of this book series would like to express the utmost gratitude and thanks to all reviewers in the technical team for making this volume a success.

Seri Iskandar, Malaysia — Ir. Dr. Mokhtar Awang, PEng, CEng
Serdang, Malaysia — Ar. Meor Razali Meor M Fared

Contents

Thermal Comfort Consequence of Pavement Material at Pedestrian Level in Melaka Town

Golnoosh Manteghi, Tasneem Mostofa, Ahmad Shukri Yahaya and Hasanuddin Bin Lamit

Abstract The pavement materials used to create urban spaces and footpaths are known to significantly affect the pedestrians' comfort levels. Consequently, this paper is focused on the influence of the outdoor thermal comfort from multiple materials being used to construct the footpaths of Melaka Heritage Town. The hourly data pertaining to the meteorological elements and outdoor thermal comfort of the pedestrian area of Melaka Heritage Town (air temperature, relative humidity, globe temperature, and wind velocity) used in this work were collected via field experiment. The RayMan Software Package (PET) was used to elucidate the thermal environment of six scenarios, made up of multiple river widths and pavement materials. The results confirmed the bare red clay pavement materials that are close to adequately wide waterbody is the best scenario to maintain a slightly warmer comfort range at the pedestrian levels for creating suitable thermal comfort. Thermal comfort assessments at pedestrian levels can also be used to elucidate the benefits of the creation of microclimates toward improving urban designs.

Keyword Pavement material · Thermal comfort · Waterbody · Physiological Equivalent Temperature (PET) · Melaka Heritage Town

1 Introduction

Many researches are currently being conducted on the outdoor thermal comfort in the context of subtropical, temperamental, and cold climates, as per [2, 3, 5, 7, 10, 11, 14]. For the past two decades and coinciding with the general increase in urban microclimates, the number of researches pertaining to this subject has also increased. Current and ongoing changes to the urban density and street mesh of modern urbanity and the monitoring of neighborhood microclimates prevent the realization of the

G. Manteghi · T. Mostofa (✉) · A. Shukri Yahaya
Faculty of Architecture and Built Environment (FABE), Infrastructure University Kuala Lumpur (IUKL), Kajang, Malaysia
e-mail: mtasnem@gmail.com

H. B. Lamit
Fakulti of Alam Bina & Ukur, Universiti Teknologi Malaysia (UTM), Skudai, Malaysia

M. Awang and M. R. Meor M Fared (eds.), *ICACE 2019*,
Lecture Notes in Civil Engineering 59, https://doi.org/10.1007/978-981-15-1193-6_1

best human thermal comfort. Furthermore, randomized urban design resulted in the development of uncomfortable regions between building blocks [1, 6, 12]. Previous research posited that pavements could decrease the heat island impact via its use phase of pavements. A significant percentage of land in urban areas is covered with multiple types of pavement, encompassing parking areas, streets, plazas, footpaths, and playgrounds. It should be pointed out that an uncomfortable microclimate environment discourages people from taking a walk around an area, which would result, in the long run, in health problems due to the lack of exercise among the populace [6]. Previous works posited that research on outdoor human comfort within tropical climates remains in its infancy, and studies focusing on thermal comfort at tourist attraction sites are even more limited. Earlier works on this topic reported the influence of high albedo materials on outdoor/indoor environments, where most elucidated the application of these materials on roofs. However, not many works focused on the effect of pavement on the ground, which is adjacent to the pedestrian levels, upon thermal comfort. Malaysia, where tourism is a source of revenue and outdoor activity is expected in most places of attractions, thus thermal comfort is [9]. This paper used a busy tourist pedestrian footpath in Melaka Heritage Town constructed with different pavement materials. This paper intends to elucidate the thermal components of the environments in tropical Malaysia.

2 Research Method

A quantitative pilot test was used to quantitatively elucidate outdoor thermal comfort, where the pilot test focused on determining the thermal conditions of outdoor spaces at pedestrian levels of Melaka Heritage Town (Fig. 1).

It was conducted during regular days from July 28, 2018 to June 4, 2018 from 10 am to 5 pm. The first phase involves collecting principal data via information from the site, encompassing observation (i.e., walkthrough), meteorological data, reports, material, and electronic documents. On the back of this data, site conditions were

Fig. 1 Bird's-eye view of the study area

Table 1 Description of the scenarios

Scenarios	Pavement material	River width (m)	Distance from waterbody (m)	Influential parameter	Selection
Scenario I	Brick tile	18	0–3	Trees	x
Scenario II	Bare red clay tile	18	0–3	Nominal vegetation	√
Scenario III	Brick tile	15	30–35	N/A	√
Scenario IV	Brick tile	15	0–3	N/A	√
Scenario V	Bare red clay tile	15	0–3	N/A	√
Scenario VI	Bare red clay tile	18	30–35	N/A	√
Scenario VII	Bare red clay tile	15	30–35	N/A	√

generalized and categorized into seven scenarios, as per the context of the current situation and planning, the width of the waterbody, pedestrian pavement material, and distance from the waterbody. During the first phase, specific information on all the scenarios was gathered. With the help of that information, field measurements were conducted, which was the second phase of this work. The third phase involves concentrating on a suitable outdoor thermal comfort index, also called Physiologically Equivalent Temperature (PET), for the evaluation of the conditions of human thermal comfort in multiple outdoor spaces [4]. The influences and insights were collected and reported in the form of guidelines for the improvement of thermal comfort at pedestrian levels in the context of Melaka Heritage Town.

2.1 *Filed Measurement Scenarios*

The pilot test outlined seven scenarios on the base of existing situation of the waterbody and current setting of the footpaths of the town. However, in the interest of an accurate quantification of the influence of pavement material with waterbody, one scenario, pertaining to vegetation, was omitted. These scenarios are tabulated in Table 1.

2.2 *Estimation of Outdoor Thermal Comfort*

RayMan 1.2 was utilized in this work to determine the levels of thermal comfort. The software package does this using personal (height, weight, age, sex, clothing, and activity) and climatological (air temperature, wind speed, relative humidity, and

Table 2 Performance of the measuring instrument

Instrument name	Parameters	Range	Accuracy	Resolution
GM816A digital anemometer	Va (ms^{-1})	0–30 ms^{-1}	±5%	0.01
RC-4HC data logger	Ta (°C), RH (%)	−30–+60 °C 0–99%	±0.18 °C ± 2.5 %s	0.02 °C 0.03%
General handy heat index	Tg (°C)	0 °C–50 °C	±1 °C	0.1

mean radiant temperature) data. Table 3 tabulates the input data utilized to determine the thermal comfort at pedestrian levels at Melaka Heritage Town. There is a small and almost negligible PET variance between men and women of standard height and weight (~0.1 °C). Therefore, this work selected males who are 1.75 m in height, weighing 70 kg, and are 35 years old as its calculation standard. Statistics tells us that the activity of 87% of tested peoples on the pedestrian side walkway involves walking (~0.9 m/s), which corresponds to a metabolic rate of 115 W/m^2 (taken as the calculation parameter). The mean clothing thermal resistance of the tested persons is ~0.90 clo.

3 Results

3.1 The Microclimate of the Selected Scenarios

In the pilot test, the annual weather was used in order to accommodate variations such as rain, cloudy, and sunny bright days in the analyses. Meteorological determinations of air temperature, relative humidity, wind velocity, and globe temperature were recorded on each survey day. Table 2 details the instruments' used for this purpose. The survey began at 10 am due to the position of the sun and end at 5 pm. Air temperature was between 29 and 42.7 °C. The highest relative humidity was cloudy, at ~75.9%, while the lowest was at noon, at 45%, averaging to ~60% for the day. The maximum outdoor wind velocity was 2.8 m/s, with its minimum at 0 m/s. The average air velocity on each survey day was 1.5 m/s, and at more than 68% of the time, it was 1.5 m/s. The highest global temperature was 44.6 °C, while the lowest was 32.7 °C. The meteorological data is summarized in Table 3.

3.2 Thermal Comfort in Selected Scenarios

The pilot test intended to compare the influence of the PET values of multiple scenarios within the footpaths of Melaka Heritage Town. The minimum PET value in the case of all of the scenarios was during the start of the survey and late afternoons. It can be seen in Figs. 2, 3 and 4 that during data collection, the PET values mostly exceeded

Table 3 Average of meteorological measurement during the survey

Scenarios	Average				
	Ta (°C)	RH (%)	V (ms^{-1})	Tmrt (°C)	PET (°C)
Scenario II	35.43	54.91	0.88	36.31	33.76
Scenario III	31.08	69.37	0.85	44.26	34.92
Scenario IV	33.07	60.01	1.11	53.36	38.05
Scenario V	31.69	64.80	1.58	70.88	36.11
Scenario VI	31.9	62.21	0.68	44.91	36.23
Scenario VII	33.62	55.9	0.7	71	41.46

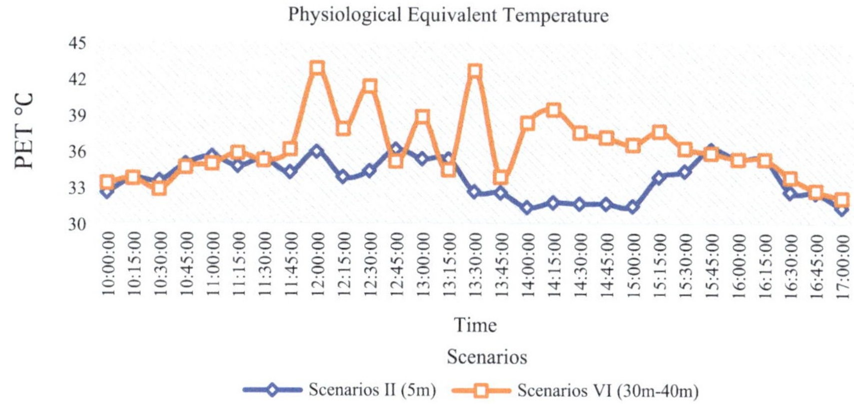

Fig. 2 PET in Scenario II and Scenario VI

the upper comfort range limit (26–30) in the case of a tropical climate [8]. Between 13:40 and 16:20 pm, whether the river width was 15 m or 18 m, the PET values exceeded the preset limit to levels that are beyond discomfort to actual deprivation. The values of the calculated PET in Scenario II (Fig. 2) where the pavement material is bare red clay tile and river width is 18 m were moderately significant than that of Scenarios III, VI, V, VI, and VII during the measurements. During measurements, especially at noon, it is noticeable that the PET values of scenario III were closer to the warm range slightly, even though it is physically located far from the Melaka River with 15 m width condition. And moreover, sporting the same brick pavement in the case of its pedestrian footpath, while lacking any operative screening from its surrounding neighborhoods because of the cloudy weather. Scenario IV, which was closer to Melaka River with 15 m of width, sports a brick pavement for its pedestrian footpath, while also lacking any effective screening due to its surroundings. It was near to the hot range at all times, as per this study. The values of the calculated PET in Scenarios V and VI were under warm range than that of other scenarios during the measurements. From Fig. 4, Scenario VII's graph and average PET from

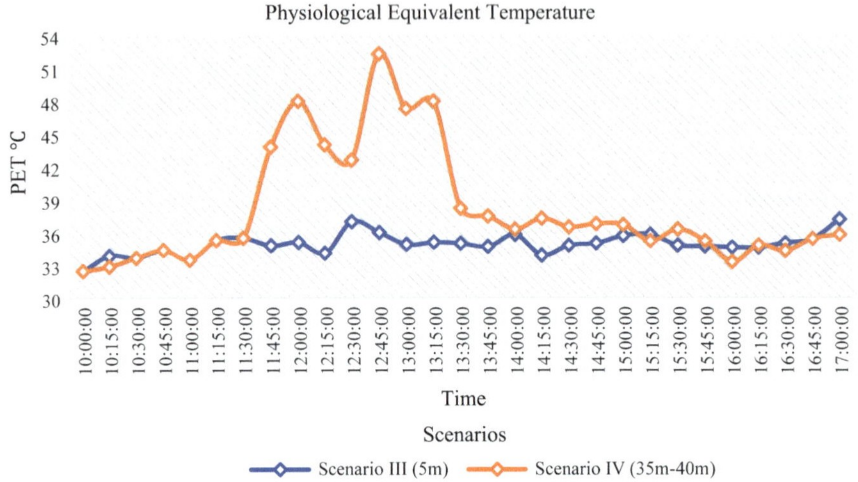

Fig. 3 PET in Scenario III and Scenario IV

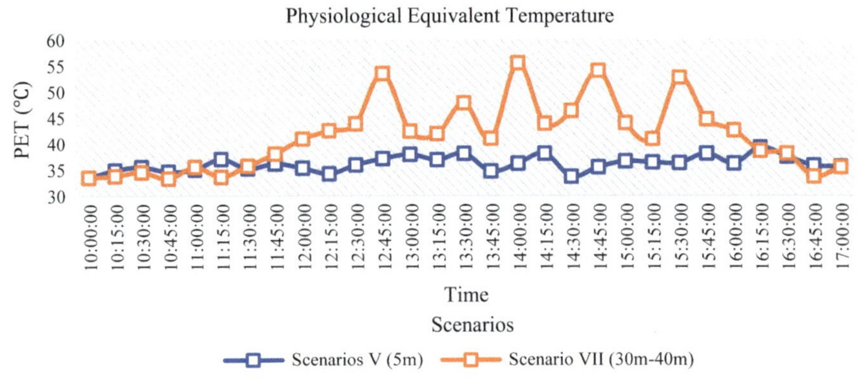

Fig. 4 PET in Scenario V and Scenario VII

Table 3 show that it went up to 41.46 °C, implying a very hot weather from 12 pm to 16:30 pm.

4 Discussion

Field measurements of the pedestrian footpath confirmed that light, laid bare on the red clay tile pavement, renders the microclimate of the area more thermally comfortable during daytime. We utilized six scenarios in the context of the pavement

The Opportunities and Challenges of Using Parametric Architectural Design Tools to Design with Full-Culm Bamboo

Case Study: A Design for a Hyperbolic Paraboloid for Gutter-Less Rainwater Capture Using Full-Culm Bamboo

John Osmond Naylor

Abstract Tropical developing economies are some of the most vulnerable societies to natural disasters, and by 2050, some 50% of the world's population will live in the tropics. Tropical developing economies already have a shortage of housing which lacks structural quality, durability and is considered non-adequate. Tropical developing economies have an opportunity to utilise locally sourced lightweight natural materials such as bamboo. Computational design processes save time, which allows greater scrutiny of design options and the testing of various iterations. A challenge is in the use of computational design tools with their great accuracy, and the natural variability of full-culm bamboo. Architects will need to develop a synthesis between their current computational design processes and materials with natural variability such as bamboo in order to improve affordability, efficiency and ensure durability. This paper presents such a synthesis, and discusses a case study of an algorithm to generate a design for a hyperbolic paraboloid. This studies the capabilities of commonly used architectural design software, and observes the efficiencies and limitations of this process. The process embeds principles in the design which will increase the durability and buildability of full-culm bamboo. If we can develop these syntheses, as designers we can obtain tools which can increase the use of renewable natural materials with variability such as bamboo, and begin to meet the need for durable, functional and adequate housing in tropical developing economies.

Keywords Full-culm bamboo · Parametric design · Adequate housing · Sustainable development

---

J. O. Naylor (✉)
School of Architecture, Planning and Landscape, Newcastle University, Newcastle upon Tyne, UK
e-mail: j.naylor73@newcastle.ac.uk

M. Awang and M. R. Meor M Fared (eds.), *ICACE 2019*,
Lecture Notes in Civil Engineering 59, https://doi.org/10.1007/978-981-15-1193-6_2

1 Introduction

Tropical developing economies are some of the most vulnerable societies to natural disasters and by 2050, some 50% of the world's population will live in the tropics [1]. Over one-quarter of the urban population of South East Asian tropical developing economies reside in non-adequate housing [2–4]. The UN Sustainable Development Goal 11 (SDG11) targets by 2030 the access for all to adequate, safe and affordable housing, and suggests that the building of sustainable and resilient buildings utilising local materials should be a catalyst for development [5]. Cement production is the third-largest source of anthropogenic emissions of CO_2 and could rise by 23% by 2050 given current trends [6]. Sand for construction is also being unsustainably sourced which in the coming decades will affect the concrete supply chain [7, 8]. We need to look at new sustainable, locally available natural materials for construction, and architects will need to respond to this challenge and develop new processes to work with natural materials which ensure structural integrity and affordability. Tropical developing economies are large producers of bamboo [9], a material with good tensile and compressive properties and a low carbon footprint when sourced locally [10]. Bamboo can be worked with simple tools and can be grown locally on a village scale or even a family scale [11]. Bamboo can also absorb CO_2 and stabilise slopes to tackle the effects of deforestation [12]. If we are to increase the use of renewable materials, then we should look to non- or marginally engineered building materials to ensure that the most affordable form of bamboo, 'full-culm bamboo', (also named 'round bamboo') is used [13, 14] (Fig. 1b). Bamboo will degrade if not designed or built correctly. Exposure to UV light and moisture can bleach, crack and encourage fungal growth causing structural and aesthetic damage which impacts greatly the perception of bamboo in the mind of potential end users, reinforcing a notion of bamboo as temporary, or the 'poor man's timber' [15].

In tropical developing economies, it is a reality that the architect will not design the majority of housing [16]. The minimum construction materials are purchased, and design and engineering input is often unaffordable. Architects can however be

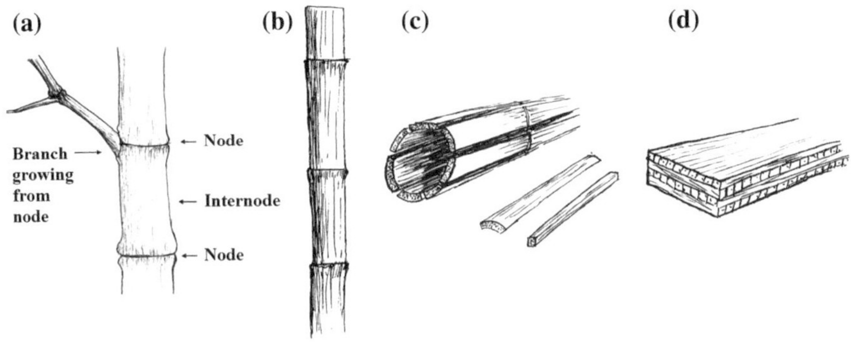

Fig. 1 **a** Culm terminology, **b** full-culm bamboo, **c** bamboo splits, **d** example of engineered bamboo

Fig. 2 Render illustrating rain collection strategy of the hyperbolic paraboloid

active in reducing the cost of design and raise awareness of good practices promoting open-source designs of adequate housing which ensure structural and aesthetic integrity [17, 18]. In order to reduce the cost of design, there have been moves by architects to develop a greater synthesis between their current computational design processes and materials with natural variability such as bamboo. The ZCB Pavilion in Hong Kong is one such example which pushes the boundaries of the design solution space of bamboo architecture [19]. Computational design tools allow the architect to visualise ideas and they can be modified and analysed interactively, though this modification can still be time-consuming. Parametric software allows us to build on this process and specify relationships among parameters and instantly output versions or iterations of a design, based on associative rules set by the designer [20]. Willis and Woodward in 2005 suggest it will be impossible to achieve a direct correlation between digital data and a constructed building. Some design parameters like material flaws, grain directions and inconsistent densities will be difficult to anticipate in modelling software. However, this gap between the building and the model will continue to narrow [21]. The following is a case study which demonstrates the use of an algorithm to generate a design for a small dwelling with a hyperbolic paraboloid roof to be built of full-culm bamboo [22]. This is a roof form which follows a convex curve about one axis and a concave curve about the other. It is practical, easily constructed from straight sections, and draws rainwater which falls on this roof, towards the two lowest points, without the additional expense of guttering (Fig. 2). This algorithm embeds certain design principles of building with full-culm bamboo, which will increase the durability, material efficiency and buildability of this roof.

2 The Tools

In this process, two pieces of software well known to the architectural profession have been used, and emblematic of a set of tools which use a visual language environment. Rhinoceros 3D [23] is a three-dimensional computer graphics and computer-aided design software which uses non-uniform rational basis splines (NURBS) to build

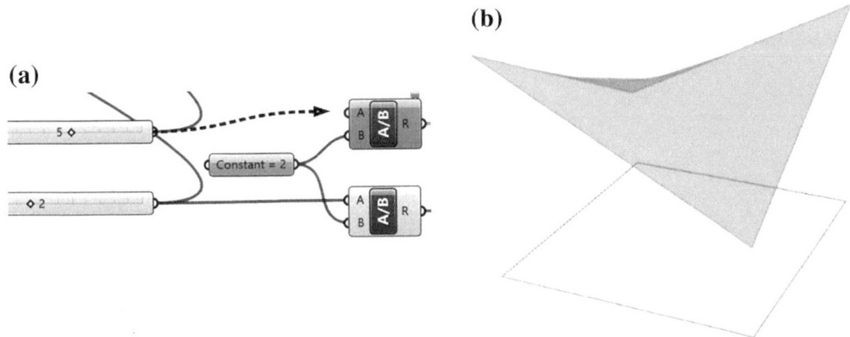

Fig. 3 **a** Example of nodes, inputs and outputs on the Grasshopper graphical algorithm editing canvas; **b** a screenshot of the Rhinoceros 3D interface showing the polygonal outline which defines the area for the surface to cover, and the hyperbolic paraboloid

geometry as opposed to a polygon mesh-based system. The second is Grasshopper [24] which is a visual programming environment which runs within the Rhinoceros 3D platform. The main interface for algorithm design in Grasshopper is the node-based editor. Algorithms are scripted by dragging components with inputs and outputs onto a canvas (Fig. 3a). A collection of components forms an algorithm, and the output of these commands is displayed in the Rhinoceros 3D window. The initial input geometry can either be assigned from Rhinoceros 3D or generated in Grasshopper. For this case study, a planar quadrilateral has been drawn in Rhinoceros 3D and assigned as the starting geometry of the algorithm in Grasshopper, though this does not have to be planar accepting the possibility of a sloping site. This algorithm will then generate a hyperbolic paraboloid above this quadrilateral, to heights defined by the user. The algorithm can respond to any quadrilateral site option to accommodate the variability of sites. Characteristics of bamboo are considered in the algorithm to improve the durability of bamboo and provide a practical means of constructing a hyperbolic paraboloid as a roof using full-culm bamboo.

3 Negating the Natural Variability of Bamboo

Bamboo has natural variability which means that there are certain characteristics of full-culm bamboo which are indeterminate. Two rational assumptions have been made in the algorithm to negate effects of variability:

- **The curvature of a bamboo culm**: The bamboo culm is modelled as a straight line, or a cylinder, as the bamboo culm for use in construction will be selected to have a very negligible curvature. It is advised that any bamboo poles selected for construction purposes will have 1% out-of-straightness limit [25].
- **The tapering diameter of a bamboo culm**: The diameter of the bamboo culm input into the algorithm is a single value across the algorithm and therefore the

design. In reality, a bamboo culm will have a diameter greater at the base than the top. This is not taken into account in this algorithm as depending on the species or specific plant the tapering can vary in extremity. When selecting the bamboo culm to be used, it is for the architect to select the culm which tapers minimally. Additionally, it is the middle section of the bamboo culm which should be used for this application. The use of the middle section will reduce the maximum and minimum diameter of the poles [26] and a decision has been made to discount this in the algorithm.

The quadrilateral building footprint is drawn in Rhinoceros 3D as a '*Closed polyline*'. It is not required to be planar which accounts for the possibility that the site may not be flat. This can reference a physical site or an isolated pre-determined quadrilateral shape defined by the needs of the brief which will consider the necessary occupancy of the building. The algorithm will generate a hyperbolic paraboloid from this initial building footprint. The algorithm for generating the hyperbolic paraboloid will extract the nodes at each corner of the quadrilateral, move the nodes vertically (Z-axis) to a point defined by the user which corresponds to the intended peak and eave heights of the roof and finally the '*Surface from 4 points*' component generates a hyperbolic paraboloid (Fig. 3b).

The roof will be constructed from straight bamboo poles with a series of longitudinal (u) poles and a series of latitudinal (v) poles placed above (Fig. 5). Within the algorithm, a new hyperbolic paraboloid is created from the generated surface by using the '*Offset Surface*' component. The distance of this offset is input into the algorithm by the user and is the diameter of the bamboo poles to be used in construction (d). Now there are two surfaces, a lower and an upper surface. Longitudinal members (u) are extracted from the lower surface and latitudinal members (v) are extracted from the upper surface (Fig. 4a). The offset of the roof is connected to the diameter of the bamboo (d) to be used. This will save a lot of time for the designer as this algorithm will adapt the design if alternative bamboo species, alternative sites or new suppliers of bamboo are required.

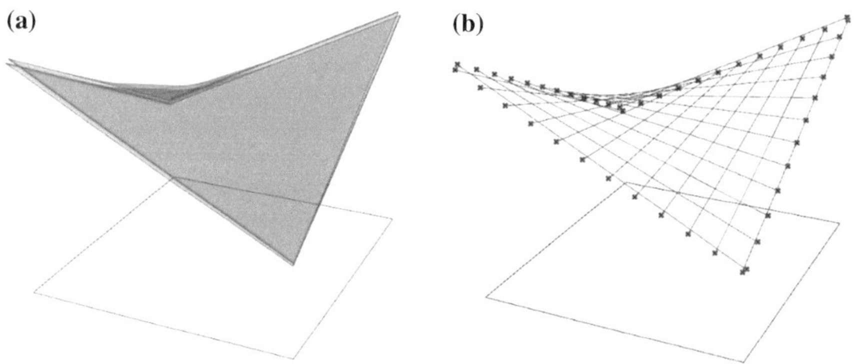

Fig. 4 **a** Two surfaces, with the upper surface offset from the lower surface, with required edge curves highlighted; **b** the 600 mm grid, with chosen node locations represented as points

4 Designing for Durability and Buildability

The distance between each pole (k) in the roof grid is again set by the user which can be determined by factors such as standardised widths of cladding materials or building codes of the region. In this example, (k) is set at 600 mm. The lengths of the opposite outer edges of both the longitudinal (u) and latitudinal (v) directions are measured. The longer of these two lengths is then divided by this user input distance (k). This will give the quantity of poles required to span the roof. The poles can then be arrayed between the first and last poles with the '*Tween Curve*' component (Fig. 4b). This ensures that the distance between the poles in the roof grid will always be less than or equal to this value (k), even if the quadrilateral plan is not orthogonal. The algorithm can instantly update the design if the grid spacing needs to be altered.

The first and last poles of each series must align in plan with an edge of the quadrilateral building footprint, so these poles can be attached to the top of the walls, beams or columns. When attaching the bamboo poles, it is important to use bolts for joints [27]. Each longitudinal (u) pole used in the roof structure will be bolted to the latitudinal (v) poles placed above. The first and last bolts in each pole which connect to a beam, or top of a wall, will need to respect the position of the nodes and be bolted on the internal side of a node (Fig. 5). It is important to place the connection in a bamboo structure in such a way that a connection is made either at a node or as near to a node as possible [11]. The algorithm projects a point on each longitudinal (u) and latitudinal (v) line in the roof grid which represents the pole, aligned with the edge of the quadrilateral footprint. The algorithm then translates this point outwards at either end of the line by a numeric value set by the user (Fig. 4b). This distance represents the desired position of the node relative to the desired bolt location. This should be roughly 2.5–5 cm. Each longitudinal (u) and latitudinal (v) member in the roof now has two points at either end of the line representing the first and last node placements. A measurement can be taken in the algorithm. This will tell the user the required distance between the first and last nodes which will be useful information when selecting the specific poles to use for each roof member.

The hyperbolic paraboloid covers the plan of the initial outline (Fig. 3b). A major issue for durability is that bamboo must be kept out of sunlight, and excess moisture and rain must be avoided [28]. The roof has an important role to play in protecting the bamboo used in the structure from driving rain and a large roof overhang is often

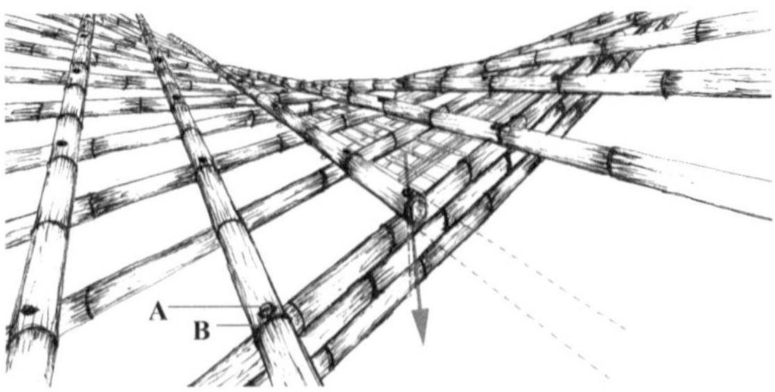

Fig. 5 Locations of bolt connections (**A**) to the beam, adjacent to a node (**B**) on the internal side

required to shelter the walls and structure. The algorithm is designed to extend each line in the roof grid, by a length which is the height above the ground plane of the roof at that location, divided by a constant value (c) (Fig. 6). The constant value (c) defines this proportion. This is a balance between the maximum angle of the driving rain and the likely exposure the overhang will have to tropical cyclones for that site, following input from a structural engineer.

Once the lines representing the roof members have been extended (Fig. 7a), the lengths of each pole can be measured (Fig. 8). This gives the user the ability to review this information against the material availability and logistical practicalities. Using

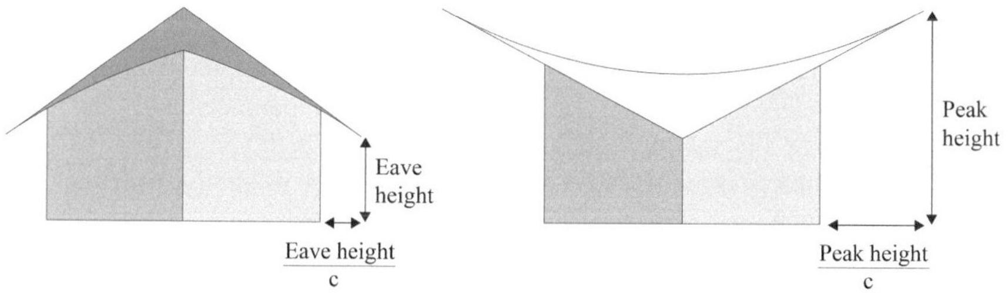

Fig. 6 The relationship between the peak and eave heights, and the roof overhang

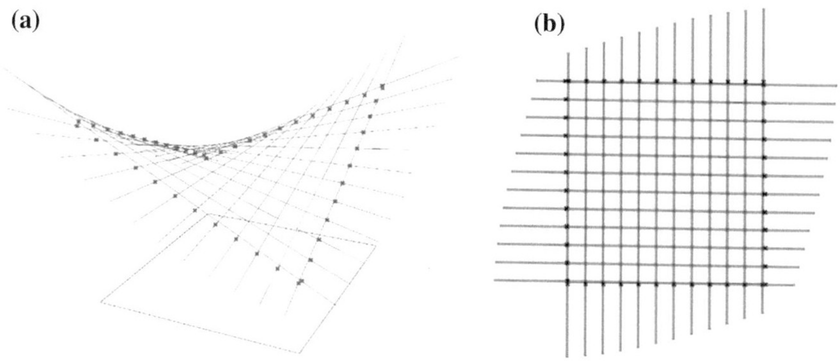

Fig. 7 The model of the roof grid: **a** as lines in perspective view; **b** as 3D volumes to represent the pole diameter in plan

Fig. 8 Grasshopper algorithm screenshot showing: lengths in cm of all the poles and lengths between the first and last nodes to be bolted on each member, in the latitudinal (v) direction

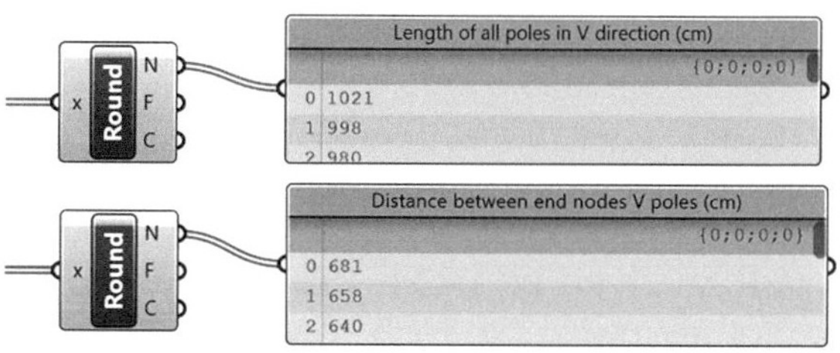

the '*Pipe*' component, the line which has represented the pole to this stage in the algorithm will generate a 3D volume which visually represents how the pole will appear (Fig. 7b). The input value of the diameter (d) is linked to the input which as discussed in Sect. 3 is also the value which was used as the offset of the roof grid. Therefore, when the user alters the value of the diameter of bamboo (d) to be used, the offset of the latitudinal (v) members will also raise simultaneously.

The output from the algorithm includes the following:

- The quantity of poles required to generate the surface.
- The lengths of bamboo poles, rounded to the nearest cm (Fig. 8). This can assist the designer or builder in determining the availability of bamboo culms, the limits of transportation to site and the cut lengths.
- The required distances between first and last nodes for each pole, which will be useful when selecting poles to be used for specific roof members.
- Visual output of the massing in order to view the form in 3D.
- NURBS geometry output which can be used to produce diagrams, drawings and rapid prototyping.
- Locations of connections where bolts may be placed. These measurements can then be used to mark out the poles prior to construction to prevent having to drill in situ which can be more dangerous to construction teams, than a workshop environment.

5 Discussion

The opportunities that processes such as these provide, allow architects to interact more closely between their computational design tools and materials with natural variability such as bamboo. However, there will always be a gap between digital and real-world environments [21]. The challenges of these tools and this process are seen in the negation of properties of bamboo. Characteristics such as the tapering of the diameter and natural curvature of the bamboo culm could have structural and architectural significance which sadly is not taken advantage of through this process. Future developments in this field can find ways to include these as well as the input of the mechanical properties of bamboo within Grasshopper through live physics and parametric structural engineering plug-ins to perform preliminary structural analysis. Overseen by a structural engineer, this can create interactive simulation, form-finding, and can further optimise the design to increase resilience to tropical cyclones. This presents the opportunity to also optimise the quantity of bamboo required for structural performance, and can also enfranchise species of bamboo currently available but unused for construction. This would however require the mechanical testing on these species to gain the input data. Questions then arise into the variability of the structural properties of bamboo even within a species, and how these variabilities are also considered in a computational design process. Given the lack of competence or literacy available in construction industries in developing economies, communicating accuracy on which the aesthetic appeal, long-term durability and structural integrity may depend can be problematic at best. A further challenge of this process is in the translation from design to construction, and therefore such a computational process may need to add robustness or margin of error into the design to compensate.

6 Conclusion

The efficiencies of the process are numerous. The algorithm embraces many quadrilateral plot shapes, and instantly amends the roof design to improve durability, simultaneously updating material lengths and quantities allowing instant evaluation against practical constraints such as material availability and budget. Processes such as these can give architects an ability to improve the durability of bamboo in their designs and save time and money for those who need resilient, sustainable buildings. If we are to succeed in reducing the global population living in non-adequate housing and achieve SDG11 by 2030 to provide access for all to, adequate, safe and affordable housing, then architects will need to more greatly align their current tools, to design with, and be vocal activists for sustainable, locally sourced, natural materials.

Acknowledgements Thank you to Osmond and Christine Naylor, Jane Wynne, Franck Vendryes, Doria Reyes Córdova, Ben Bridgens and Thomas Wortmann who provided insight and expertise, although interpretations, conclusions and any errors in this paper are my own.

References

1. Edelman A, Gedling A, Konovalov E, et al (2014) State of the tropics 2014 report 43:146–149. https://doi.org/10.1177/0011000014564251
2. United Nations (2015) The millennium development goals report 2015
3. UN-Habitat (2015) Issue paper on informal settlements. Habitat III issue Pap 22
4. UN Habitat (2018) Slum Almanac 2015/2016
5. UN-Habitat (2016) Sustainable development goals: goal 11 monitoring framework:56. https://doi.org/10.1016/j.jasms.2004.03.003
6. Harvey C (2018) Cement producers are developing a plan to reduce CO_2 emissions. E&E News, Sci Am:1–13
7. Andrew RM (2017) Global CO_2 emissions from cement production. Earth Syst Sci Data:1–52
8. Torres A, Brandt J, Lear K, Liu J (2017) A looming tragedy of the sand commons. Sci (80-) 357:970–971. https://doi.org/10.1126/science.aao0503
9. Lobovikov M, Paudel S, Piazza M, et al (2007) World bamboo resources: a thematic study prepared in the framework of the global forest resources assessment 2005
10. Van Der Lugt P, Vogtländer JG, Van Der Vegte JH, et al (2014) Environmental assessment of industrial bamboo products—life cycle assessment and carbon sequestration. In: 10th World Bamboo Congress
11. Janssen JJA (1995) Building with Bamboo: a handbook

12. Tardio G, Mickovski SB, Stokes A, Devkota S (2017) Bamboo structures as a resilient erosion control measure. Proc Inst Civ Eng—Forensic Eng 170:72–83. https://doi.org/10.1680/jfoen.16.00033
13. Paraskeva T, Pradhan NPN, Stoura CD, Dimitrakopoulos EG (2019) Monotonic loading testing and characterization of new multi-full-culm bamboo to steel connections. Constr Build Mater 201:473–483. https://doi.org/10.1016/j.conbuildmat.2018.12.198
14. Harries KA, Sharma B, Richard M (2012) Structural use of full culm bamboo: the path to standardization. Int J Archit Eng Constr 1:66–75. https://doi.org/10.7492/ijaec.2012.008
15. INBAR (2003) Proceedings of Bamboo housing workshop. In: International Network for Bamboo and Rattan (INBAR) & Bamboo and Rattan Development Program (BARADEP). WITC, Kumasi Ghana
16. Hardoy JE, Satterthwaite D (1989) Squatter citizen: life in the urban third world. Earthscan Publications, London
17. Seksan N (2017) Seksan design: open source drawings [Internet]. https://www.seksan.com/open-source-drawings
18. Aravena A (2019) Elemental [Internet]. http://www.elementalchile.cl/en/
19. Crolla K (2017) Building indeterminacy modelling—the 'ZCB Bamboo Pavilion' as a case study on nonstandard construction from natural materials. Vis Eng 5. https://doi.org/10.1186/s40327-017-0051-4
20. Jabi W (2012) Parametric design for architecture. Laurence King Publishing, London
21. Willis D, Woodward T (2005) Diminishing difficulty: mass customisation and the digital production of architecture. Harvard Des Mag:71–83
22. Weisstein EW (2019) Doubly ruled surface. In: MathWorld. http://mathworld.wolfram.com/HyperbolicParaboloid.html
23. Robert McNeel & Associates (2019) Rhinoceros 3D
24. Robert McNeel & Associates (2019) Grasshopper
25. Kaminski S, Lawrence A, Trujillo D (2016) Structural use of bamboo Part 1: introduction to bamboo. Struct Eng 94:40–43
26. Minke G (2012) Building with Bamboo. Design and Technology of a Sustainable Architecture. Birkhäuser, Berlin, Basel
27. Hidalgo-López O (1981) Manual de Construccion con Bambu. Estudios Técnicos Colombianos
28. Kaminski S, Lawrence A, Trujillo D, King C (2016) Structural use of bamboo. Part 2: Durability and preservation. Struct Eng 94:38–43

The Effect of Grouting Material to the Compressive Strength and Modulus of Elasticity of the Biaxial Interlocking Block Prisms

Fauziah Aziz, Mohd. Fadzil Arshad and Hazrina Mansor

Abstract This paper aims to investigate and evaluate the influence of the grouting material as a reinforced method and thin layer connection binder on the prism's compressive strength behavior as a load-bearing member. The test specimens consist of Biaxial Interlocking block (BIB) prisms, namely, G1 and G2. G1 is a prism sample constructed without grouting material, but with thin layer connection binder, while G2 is a prism stacked with thin layer mortar filled in with grouting material inside the grooves. A series of test, on raw materials and the masonry constituent, such as block unit, thin layer mortar, and grouting material. Based on the data obtained, results present that the G2 sample produces higher compressive strength compared to G1 at a significant level. The results also showed that modulus of elasticity (MOE) of G2 is higher than G1. As a result, it shows that grouting material plays a significant role in enhancing the compressive strength of the masonry structures.

Keywords Compressive resistance · Biaxial interlocking concrete block · Grouting prisms · Thin layer connection binder

1 Introduction

Masonry is the oldest construction method in the world with composite and complex material, in which solid, hollow blocks, or interlocking concrete block with or without mortar are used. Its simplest arrangement formed by blocks joined together by mortar or without mortar is the main reason for the success of masonry project. But, the capacity and behavior of masonry, however, are neither that of the block nor of the mortar [1].

Generally, there are several types of masonry units used in masonry construction, such as clay brick, cement brick, and concrete block. Among these, cement brick is the most popular material used in building construction. However, due to its weakness such as the compressive strength of the cement bricks, requirement of skilled worker during assembling, and increasing the time of construction, concrete block become as

F. Aziz (✉) · Mohd. F. Arshad · H. Mansor
Faculty of Civil Engineering, University Technology Mara, Shah Alam, Malaysia
e-mail: fauziah2876@gmail.com

M. Awang and M. R. Meor M Fared (eds.), *ICACE 2019*,
Lecture Notes in Civil Engineering 59, https://doi.org/10.1007/978-981-15-1193-6_3

a new alternative masonry unit for industry player. Concrete block is divided into two categories, a solid block and hollow block. Despite all these, interlocking concrete block in hollow block category, also known as dry-stack system and mortarless system, is a new innovation to replace the function of solid blocks in terms of reducing the dead load by numbers of void, but, with high compressive strength, as well as reducing the required unskilled labor, reducing the construction time, and decreasing the effects of workmanship. Dry-stack system or mortarless is a method of masonry wall construction without mortar application in bed and head joints [2–4]. This system reacts through interlock to lead the masonry unit, connected tightly, to be an entire system [5]. Therefore, this system requires to be filled in with grouting material in the groove areas to react as a load-bearing wall application.

According to [6], the grouted process that involves the combination of cement and sand is a well-known method to enhance the capacity of structural elements. It is supported by [2], in which it has been mentioned that a mortarless system should be grouted to increase its performance as a load-bearing wall. Furthermore, [7] described that grouting material is a combination of cement, aggregates, and water mixed together to fill the hollow to enhance their capability under compression load.

Besides that, [8] agreed that enhancement of the structural capacity of the wall system could be achieved by increasing the volumetric ratio of the grouted area. Furthermore, [9] in their research found out that the insertion of grouting material in the hollow concrete walls increased about 80% of the resistance capacity. In other words, the effort to improve the load-bearing capacity of the walls is a good method to enhance their capability to resist the load [10].

There are several factors influencing the mechanical properties of masonry walls such as mechanical properties of masonry unit, the thickness of bed joints and head joints, and mechanical properties of mortar joints [11]. However, in this dry-stack system, there is no mortar application in bed and head joints. The compressive strength of a masonry wall strictly depends on the masonry unit and interlocking system. The literature also indicates that mortar strength not much reflects on the masonry strength. Apart from that, to enhance the capability of the prisms, the effect of the grouted and interlocking system was also counted. Regarding [11], the compressive strength of hollow CS block masonry depends on the mortar strength in bed joint, joint thickness, form of a block, the volume of voids, and arrangement of voids.

Based on EN 1996-1-1, characteristic compressive strength f_k of concrete masonry made with thin layer mortar, in bed joints of thickness 0.5–3 mm, may be obtained from either results of tests by EN 1052-1 or equation: $f_k = K\ f_b^{0.85}$. From the equation, it has been shown that the characteristic strength of a block is most influenced by the characteristic strength of the masonry wall. Research conducted by [12] on 31 different types of blocks found out that the calculation of the characteristic compressive strength of the thin joint hollow concrete masonry made with dense or lightweight aggregates, proposed by EC6, depends only on the capacity of the block.

A part of the masonry unit strength or the strength of mortar plays an important role in developing a compressive strength of dry-stack masonry wall and the strength of grouting material, and connection binder cannot be neglected [1].

This study aims to investigate the capacity of biaxial interlocking block system as a load-bearing wall under compression load, mainly on the compressive strength of the block, the compressive strength of grouting material, and the compressive strength of thin layer connection mortar.

2 Experimental Program

2.1 Material Properties

Materials used in this study are biaxial interlocking block, sand, cement, water, and cement glue. The details of the materials used are discussed as follows.

2.2 Biaxial Interlocking Concrete Block (BIPB)

Interlocking concrete block, namely, Biaxial Interlocking Block (BIB) is a block produced and supplied by Innovative Precast Builder Sdn.Bhd. The dimensions of the BIB unit are shown in Fig. 1.

This BIB unit is 700 × 100 × 150 mm in size, which consists of seven (7) numbers of grooves. For this research, 40 units of IPB were involved in their physical and mechanical properties. The number of blocks required for the testing is given in Table 1. Out of 40 units, 20 units are grouped as a hollow unit and 20 units as a grouted.

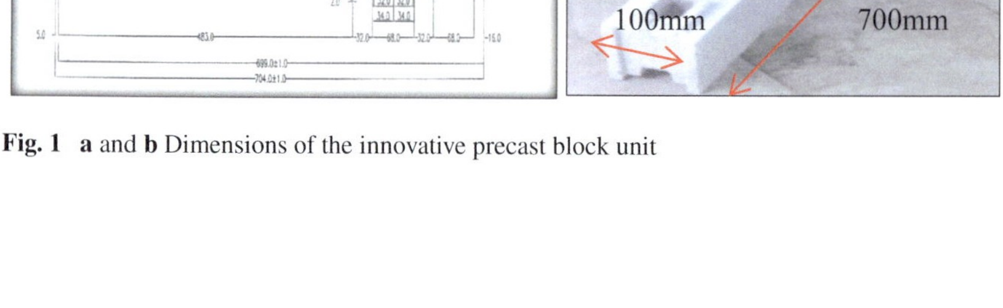

Fig. 1 **a** and **b** Dimensions of the innovative precast block unit

Table 1 Numbers of BIB unit samples

BIB sample	No. of sample	Type of testing
Hollow unit	10	Compressive strength test/modulus of elasticity
	10	Density test/dimension check
Grouted unit	10	Compressive strength test/modulus of elasticity
	10	Density test/dimension check

2.3 Grouting Material

In this study, the grouting material was used as a reinforced method to enhance the capability of the block. Besides that, to comply with the Malaysian standard requirements for a load-bearing wall during construction, the grooves should be infilled. For the grouting process, it involves the mixture of ordinary Portland cement, sand, and water. The mix proportion of this material is shown in Table 2.

Grouting material design mix was as shown above. Based on [13], the design mix is 1:3 with 0.5 water–cement ratio which produced enough compressive strength to fill in the grooves. Meanwhile, the chemical compositions of OPC Type 1 are shown in Table 3.

Table 2 Design mix of grouting material

Design mix	Cement	Aggregates	Water/cement ratio (w/c)
1:3	Ordinary Portland cement	Sand	0.5

Table 3 Chemical compositions for ordinary Portland cement

Chemical compositions	%
CaO	63.00
SiO_2	20.00
Al_2O_3	5.70
MgO	0.99
Fe_2O_3	2.90
SO_3	3.55
K_2O	3.50
LOI	2.8

Table 4 Compositions of super adhesive

Compositions	%
OPC	40
Silica sand	50
Calcium carbonate	7
Redispersible polymer powder	2
MHEC cellulose	0.5
Calcium formate	0.5

Table 5 Super adhesive properties

Type of material	Tensile adhesion strength	Shear adhesion strength
Master super adhesive for ceramic and marble	840 N	14.80 kN

Table 6 Design mix of thin layer connection binder material

Design mix	Cement/super adhesive		Water/cement ratio (w/c)
1:2 (connection binder)	(1) OPC	(2) Super adhesive	0.5

2.4 *Thin Layer Mortar*

In this research, cement glue was used as a thin layer mortar between the BIB units during the construction of prisms. The properties of super adhesive material are shown in Table 4.

Meanwhile, the mechanical properties of the super adhesive material are shown in Table 5. This product is produced by Italtech Sdn.Bhd.

Besides that, the thin layer mortar mixed design is 1:2 ratio with water–cement ratio of 0.5. One part of ordinary Portland cement Type 1 combined with two parts of super adhesive material is shown in Table 6.

2.5 *Sample Preparation*

The experimental program involved the preparation of BIB unit sample for determination of its mechanical properties and the preparation of cube sample for grouting material and thin layer mortar properties. The samples were divided into two groups as hollow and grouted, as shown in Table 7 and Fig. 2. The experimental program involved the preparation of cube sample for determination of the mechanical properties of the grouting material and thin layer mortar. The mortar mix and thin layer mortar were prepared to achieve the required strength stated in Eurocode 6 Design

Table 7 Numbers of BIB units for testing

BIB	No. of sample	Type of testing	Standard compliance
Hollow and grouted unit	20	Compressive strength test/modulus of elasticity	MS 1933-1-2017: Methods of test for masonry units—Part 1: Determination of compressive strength
	20	Density test	MS 1933-13-2007: Methods of test for masonry units—Part 13: Determination of net and gross dry density of masonry units (except for natural stone)

(a) **(b)**

Fig. 2 **a** Hollow BIPB unit, **b** grouted BIPB unit

of Masonry. Twelve (12) numbers of cubes with sample sizes of 150 × 150 × 150 and 50 × 50 × 50mm were prepared in the concrete laboratory. The samples were cured by immersing in water for 7 and 28 days of age. Meanwhile, six (6) numbers of the grouted interlocking hollow block were tested to certify the Malaysian Standard MS 1933-1-2007, compressive strength of the unit. Blocks were cut and ground for surface preparation and capped on top and bottom to reproduce the same contact area to simulate the actual compression state of the block in the masonry. The evaluation of the grouting material and thin layer mortar compressive strength was carried out based on BS EN 1015-11.

Meanwhile, for BIB prisms, the total numbers of six (6) prisms were prepared for testing. There are hollow prisms and grouted prisms. The details of the prisms are shown in Table 8. The masonry prisms were constructed with interlocking concrete units of seven grooves in the stack bond and prism's production steps are shown in Fig. 3. Face shell bedded joint was used at bed joint layer. The connection binder thickness is about 2 mm for the bed joint. After 24 h, three of the prisms were concrete

Table 8 Sample preparation for prisms

Size of the sample (mm)	Sample	No. of sample	Standard compliance
700 × 100 × 450	Hollow prisms + thin layer mortar (G1)	3	MS 1934-1:2007 (confirmed 2014) Method of test for masonry-part 1: Determination of compressive strength
	Hollow prisms + grouted material + thin layer mortar (G2)	3	

(a) (b) (c)

Fig. 3 **a**, **b**, and **c** Preparation of prism samples

infilled. All the prisms were kept immobile until the age of 28 days, protected from the sun and wind.

2.6 Testing Procedures

The testing procedures involved grouting material, thin layer connection binder, BIB unit, and BIB prisms in compliance with Malaysian standard and British standard.

2.6.1 Compressive Strength and Density of Grouting Material

The compressive strength of mortar was determined by using 50 mm cubes as per ASTM C109/C109M—Standard Test Method for Compressive Strength of Hydraulic Cement Mortars. The test was conducted for six (6) cubes and reported the average value as the test result for both 7-day and 28-day compressive strength.

2.6.2 Compressive Strength of Thin Layer Mortar

The compressive strength of thin layer connection binder is determined by using 50 mm cubes as per ASTM C91/C91M—Standard Specification for masonry cement together with ASTM C109/C109M—Standard Test Method for Compressive Strength of Hydraulic Cement Mortars. The test was conducted for six (6) cubes and reported the average value as the test result for both 7-day and 28-day compressive strength.

2.6.3 Density Test of BIB Unit

All the samples will follow the MS 1933-13-2007: Methods of test for masonry units—Part 13: Determination of net and gross dry density of masonry units (except for natural stone). A total of 20 units of BIBs as hollow or grouted were used to determine the net and gross density of masonry units. For whole unit specimens, specimens were dried to constant mass, in a ventilated oven at a temperature of 70 ± 5 °C. Constant mass is reached, if during the drying process in two subsequent weighing with a 24-h interval, the loss in mass between the two determinations is not more than 0.2% of the total mass.

2.6.4 Compressive Strength Test of BIB Unit

All the samples will follow MS 1934-1:2007(confirmed 2014) method of test for masonry—part 1: Determination of compressive strength. A minimum number of samples required are six (6) per 1000 consignment. First, immerse the specimens in water at a temperature of 20 ± 5 °C for a minimum period of 15 h and subsequently allow them to drain for 15–20 min. According to clause 7.4.2, where the net loaded area of masonry units with a frog, which is intended to be filled with mortar in practice (see also 7.2.3), is not less than 35% of the gross area, then the compressive strength shall be calculated on the basis of the net loaded surface of the frogged bed face. Where the net loaded area of masonry units with a frog is less than 35% of the gross area, then the compressive strength shall be calculated on the basis of the gross area of the masonry unit.

2.6.5 Compressive Strength Test of Prisms

The scope of the masonry evaluation on prisms is to investigate and determine the characteristic compressive strength (f_k) according to European Standard test method stated in BS EN 1052-1. In this study, three masonry prisms were constructed for each case, i.e., interlocking each unit with thin layer mortar at their bed and head joints and putting a grouting material as a reinforced method. These tests were performed to evaluate and better understand the behavior of the prisms subjected to the static compressive load. The prisms were tested after 28 days of their construction. A metallic beam was installed on top of the prism's surfaces to distribute the force applied to the uniformly distributed load. The load is applied constantly at 0.08 mm/s and directly measured by the load cells of the press. Displacement transducer strain gages with 100 mm of rated capacity were fixed on the center of the sample. They were able to measure the displacement in the vertical and horizontal directions. The modulus of elasticity of each prism was calculated in the range corresponding to the secant curve between 5 and 30% of its actual compressive strength.

2.7 *Determination of* $\mathbf{f_k}$ *Value and Modulus of Elasticity (MOE)*

The characteristic compressive strength of masonry f_k then was determined according to EN 1052-1 as a greater value between the following:

- min $(f/1.2; f_{\min})$,
- 5% fractile value based on a confidence interval of 95%,

$$f_i = \frac{F_{i,max}}{Ai} \tag{1}$$

where $f_{\min}$ is the lowest individual result of the three (3) prism specimens and f is the mean strength of the three (3) prism specimens.

Meanwhile, MOE value can be calculated using the formula given below:

$$E_i = \frac{F_{i,max}}{3x\varepsilon ixAi} \tag{2}$$

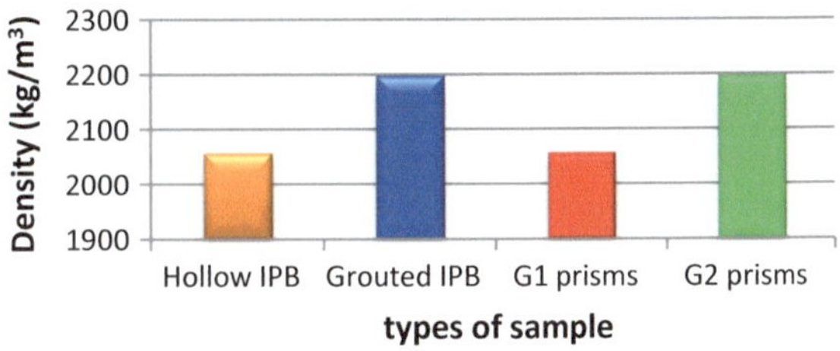

Fig. 4 Comparison of the density value for BIPB blocks and prisms

3 Results and Discussion

3.1 Density Value of the Sample

Figure 4 presents the density value of hollow BIB units, grouted BIB units, G1 prisms, and G2 prisms. From the figure, it is shown that the density value of hollow BIB units is 2059 kg/m^3, grouted BIB units is 2200 kg/m^3, G1 prisms is 2059 kg/m^3, while G2 prisms is 2200 kg/m^3.

From the figure, it is found that the density of grouted BIB units and prisms is higher than the others. The higher value density of grouted BIB units and prisms might be due to the effect of grouting material that fills in the groove area of the BIB units. It is also regarding the mix ratio used. Based on [14], the solid block's weights are higher than those of hollow blocks which lead to higher density. Besides that, [15] found out that the decrease in weight can be explained by the Law of Mixtures. The weight of a sample influences the density of the sample as found out by [16]. As a conclusion, the density of the sample will be influenced by the future application of the block (Table 9).

3.2 Compressive Strength of the Prisms Sample

Figure 5a, b presents the compressive strength of hollow BIB, grouted BIB, G1 prisms, and G2 prisms. From the figure, it is shown that the compressive strength of a hollow BIB is 19.8 N/mm^2, grouted BIB is 23.7 N/mm^2, G1 prisms is 9.7 N/mm^2, while G2 is 10.6 N/mm^2. From the figure, it is found that the compressive strength of grouted BIB is higher than hollow BIB unit. Meanwhile, G1 prisms produced lower compressive resistance than G2 sample. Regarding [17], their study showed that the grout bond strength is improved with the decrease of the water/cement ratio. The compressive strength of the grouted BIB is higher than the hollow unit and might be due to the effect of grouting material compressive strength when reacting with the hollow block. A study conducted by [18] mentioned that the voids number of a block unit significantly affects the compressive strength value.

Table 9 The compressive strength of the hollow and grouted prisms

	Compressive strength (N/mm^2), f_i	The characteristic compressive strength of masonry (f_k)
		Experimental
Hollow prisms (G1)		9.68
G1-1(f_i)	11.84	
G1-2(f_i)	12.02	
G1-3(f_i)	11.00	
$f_{i,\ min}$	11.00	
f	11.62	
Grouted prisms (G2)		10.6
G2-1(f_i)	14.10	
G2-2(f_i)	12.00	
G3-3(f_i)	12.10	
$f_{i,\ min}$	12.00	
f	12.72	
MOE (E_i)		
Hollow prisms	11700 N/mm^2	
Grouted prisms	16509 N/mm^2	

The results of the characteristic compressive strength (f_k) of the tested prisms have shown the reduction between G1 prisms and G2 prisms. It is due to the effect of grouting material as an enhancement agent of the compressive strength of the block unit. Based on [7], it was found that the compressive strength of grout/mortar is related to the behavior of prisms in a linear direction. The compressive strength of the G2 prisms was higher than that of the G1 prisms due to the compressive strength of the grout, occupying 51% of the gross area, which was much higher than that of the outer shell (hollow prisms) [19]. Furthermore, the effect of a thin layer mortar at head and bed joints also plays a significant effect to the compressive resistance. The study carried out by [11] shows that the compressive resistance of hollow prisms depends not only on the mortar strength in bed joint and joint thickness but also on the form of a block, the volume of holes, and arrangement of holes. The results of the study by [7] showed the importance of achieving compatibility of the masonry components, in order to improve the performance of the masonry structure. Furthermore, [20] found that the compressive strength of the grouted prisms is increased by about 50% compared to hollow prism compressive resistance.

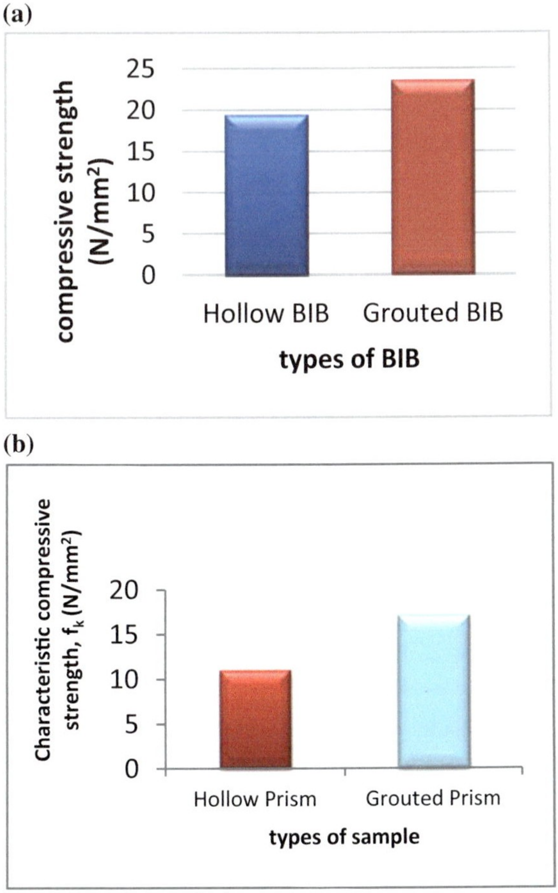

Fig. 5 **a** The compressive strength of the BIB unit at 28-day **b** the characteristic compressive resistance of the BIB prisms

3.3 *Modulus of Elasticity of the Samples*

Modulus of elasticity (MOE) of the unit was obtained by testing BIB unit under axial load, and reading by strain gage in the vertical direction, while MOE of prisms was calculated from the testing of compressive strength. The grouted BIB unit is having 23404 N/mm^2, while hollow BIB is having 20659 N/mm^2. Meanwhile, G1 prisms are having 11700 N/mm^2 and G2 prisms are having 16509 N/mm^2. From Fig. 6, it is found that grouted BIB units have higher MOE than others. The higher MOE might be due to the higher density and compressive strength of the sample found in Fig. 1. It is supported by [21] that static modulus of elasticity has linear correlation with the cube compressive strength. Besides that, [22] stated that the relationship between elastic modulus and compressive strength of masonry is linearly related. Furthermore, [23] claimed that concrete material is harder than clay brick that leads to the higher modulus of elasticity.

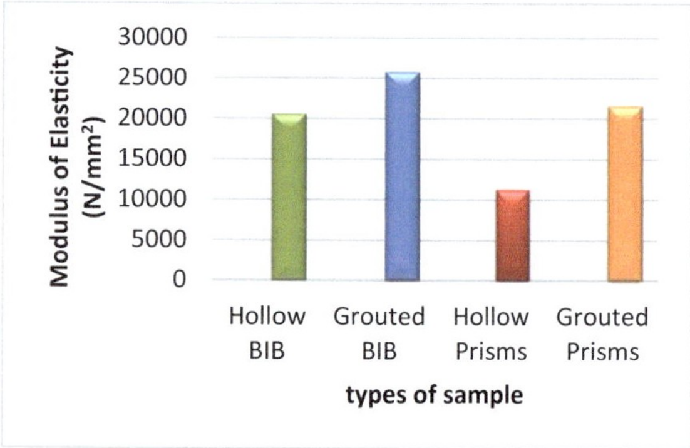

Fig. 6 Modulus of elasticity of the samples

4 Conclusion

This paper presented a comprehensive experimental program with the objective of assessing the compressive strength of hollow prisms and grouted prisms. Nominal resistance of blocks was 23.7 N/mm^2 with the grouting material produced with shrinkage-compensating admixture. Prisms's compressive strength was evaluated considering the compressive strength of blocks, the thin layer mortar bedding type, and the compressive strength of grouting material as an enhancement method. The main conclusions of the research presented here are as follows:

a. The control prisms constructed with thin layer mortar but without grouting material showed the lowest compressive resistance.
b. All the G2 prisms showed a significant increment in compressive strength compared to G1 prisms.
c. The modulus of elasticity shows a significant linear correlation with the compressive strength and density of the sample.
d. Experimental results show a significant result of increment in compressive strength of prisms due to the enhancement method compared to the dry-stack system.

Acknowledgements The authors gratefully acknowledge the Innovative Precast Builder Sdn. Bhd. as a main supplier of the interlocking concrete blocks in this research, University Technology Mara, Malaysia, and wish to thank the Ministry of Education, which provided the scholarship and also whoever was involved in this project.

References

1. Mohamad G, Fonseca FS, Vermeltfoort AT, Martens DRWW, Lourenço PB (2017) Strength, behavior, and failure mode of hollow concrete masonry constructed with mortars of different strengths. Constr Build Mater 134:489–496

2. He XX, Lin Y (2014) Research on compressive strength of grouted dry-stack concrete block masonry. Appl Mech Mater 638–640:202–207
3. Atamturktur S et al (2017) Compressive strength of dry-stacked concrete masonry unit assemblies. J Mater Civ Eng 29(2):06016020
4. Safiee NANA et al (2018) Behaviour of interlocking mortarless hollow block walls under in-plane loading. Aust J Struct Eng 7982(2):1–9
5. Uzoegbo HC, Ngowi JV (2003) Structural behaviour of dry-stack interlocking block walling systems subject to in-plane loading. Concr Bet 103:9–13
6. Luso E, Lourenço PB (2017) Bond strength characterization of commercially available grouts for masonry. Constr Build Mater 144:317–326
7. Martins ROG, Nalon GH, Alvarenga R. de CSSA, Pedroti LG, Ribeiro JCL (2018) Influence of blocks and grout on compressive strength and stiffness of concrete masonry prisms. Constr Build Mater 182(June):233–241
8. Seo SY, Jeon SM, Kim KT, Kuroki M, Kikuchi K (2015) Compression behavior of form block walls corresponding to the strength of block and grout concrete. Int J Concr Struct Mater 9(1):21–33
9. Camacho JS, Logullo BG, Parsekian GA, Soudais PRN (2015) The influence of grouting and reinforcement ratio in the concrete block masonry compressive behavior Influência do graute e da taxa de armadura no comportamento à compressão da alvenaria de blocos 8(3):341–352
10. Al-Sibahy A, Edwards R (2014) Behaviour of masonry wallettes made from a new concrete formulation under compression loads at ambient temperatures: testing and modelling. Constr Build Mater 63:271–280
11. Zavalis R, Jonaitis B, Lourenço PB (2018) Experimental investigation of the bed joint influence on mechanical properties of hollow calcium silicate block masonry. Mater Struct 51(4):85
12. Trad A, Ghanem H, Abbas N, Hamdan Z (2015) Thin layer concrete blockwork in compression: an experimental analysis. Eng Struct Technol 7(2):91–96
13. Singh SB, Munjal P, Thammishetti N (2015) Role of water/cement ratio on strength development of cement mortar. J Build Eng 4:94–100
14. Abdul Ghani AH, Fadzil Arshad M, Salehuddin NA, Mohamad Ridzuan AR (2015) Hydration kinetics and strength development of waste paper sludge ash (WPSA) with different particle size as binary blended cement. Appl Mech Mater 754–755: 382–388
15. Lasco J, Madlangbayan M, Sundo M (2017) Compressive strength and bulk density of concrete hollow blocks (CHB) with polypropylene (PP) pellets as partial replacement for sand. Civ Eng J 3(10):821–830
16. Raheem AA, Kayode Momoh A, Soyingbe AA (2012) Comparative analysis of sandcrete hollow blocks and laterite interlocking blocks as walling elements. Int J Sustain Constr Eng Technol 3:2180–3242
17. Izquierdo OS, Correa MR, Soto I (2017) Study of the block/grout interface in concrete and clay block masonry structures Estudo da interface bloco/graute em alvenaria de blocos. IBRACON Struct Mater J 10(4):924–936
18. Sathiparan N, Anusari MKN, Samindika NN (2014) Effect of void area on hollow cement masonry mechanical performance. Arab J Sci Eng 39(11):7569–7576
19. Hamid A, Bolhassani M, Moon FL (2013) Mechanical properties of ungrouted and grouted concrete masonry assemblages, no. June
20. Camacho J, Logullo B, Parsekian G, Soudais PR (2015) The influence of grouting and reinforcement ratio in the concrete block masonry compressive behavior. Ibracon Struct Mater J 8(3):341–364
21. Umoh AA, Olusola KO (2012) Compressive strength and static modulus of elasticity of periwinkle shell ash blended cement concrete. Int J Sustain Constr Eng Technol 3(2):2180–3242
22. Zhou Q, Wang F, Zhu F, Yang X (2017) Stress–strain model for hollow concrete block masonry under uniaxial compression. Mater Struct Constr 50(2):106
23. Radovanovic Z, Grebovic R, Dimovska S, Serdar N, Vatin N, Murgul V (2015) The mechanical properties of masonry walls—analysis of the test results. Procedia Eng 117:865–873

Effect of Soil Hydraulic Properties on Infiltration of Rainwater into Unsaturated Soil

Aniza Ibrahim and Muhammad Mukhlisin

Abstract Infiltration of rainwater is a major reason for soil instability in Malaysia. Rainfall event, especially in a tropical country is considered to have high intensity and long duration which will change parameters such as pore water pressure and moisture content. The rate of infiltration depends on the basic soil properties and hydraulic properties. In this paper, the effect of soil hydraulic properties on the infiltration of rainwater is discussed. There are two types of soil used in this study, sandy soil and kaolin. Hydraulic properties such as hydraulic conductivity and soil water characteristic curve (SWCC) are defined for both types of soil. Then soil column apparatus with applied rainfall to the column is used to determine moisture content using time-domain reflectometry (TDR). The result shows that kaolin has more ability to withstand the water than sand. Kaolin soil has lower hydraulic conductivity value compared to sand soil. Other than that, volumetric moisture content value is increased when water flow is applied to the soil and decreased when flow is stopped, or during drying process.

Keyword Infiltration · Soil hydraulic properties · Soil column · Unsaturated soil · Rainfall intensity

1 Introduction

Rainfall is the main cause of soil instability in Malaysia. As a tropical country, Malaysia has prolonged and high intensity of rainfall which causes the reduction of soil matric suction [2, 4]. In addition, the rainfall will increase moisture content of the soil and affect the behavior of soil. Furthermore, the hydrological changes of precipitation, infiltration, evaporation, and transpiration will also definitely change soil

A. Ibrahim (✉)
Department of Civil Engineering, National Defense University of Malaysia, Sungai Besi Camp, 57000 Kuala Lumpur, Malaysia
e-mail: aniza@upnm.edu.my

M. Mukhlisin
Polytechnic Negeri Semarang, Semarang, Indonesia

M. Awang and M. R. Meor M Fared (eds.), *ICACE 2019*,
Lecture Notes in Civil Engineering 59, https://doi.org/10.1007/978-981-15-1193-6_4

behavior [7]. Rainfall-induced soil slope failures were studied by various researchers and the main cause is the reduction of soil strength due to rainfall infiltration [6].

Understanding the hydraulic–mechanical behavior of the unsaturated soil is vital in ensuring appropriate engineering design. It is also important in applying unsaturated soil mechanics in engineering practice. There are two important parameters for hydraulic properties, hydraulic conductivity and soil water retention, which normally represents the relationship between soil water content and soil water suction.

SWCC is defined as relationship between water content and suction in unsaturated soils [5]. The parameters, such as soil suction and water content, control the permeability, volume change, deformability, and shear strength. There are two paths that can be used in obtaining SWCC: drying (desorption) and wetting (adsorption) [3]. A typical adsorption and desorption SWCC are shown in Fig. 1. The desorption of SWCC is measured by saturating the soil sample first, then applying the air pressure in a series of increments to attain different values of matric suction. The increment of pressure is applied to expel water from the specimen until an equilibrium state is reached. The water content of a soil decreases as its suction increases following the drying path or reverse process occurrence in the wetting path [1].

The permeability coefficient, *k*, is defined as the ability of the soil to drain water. Coarse grain has high hydraulic conductivity compared to fine grains when saturated. Fine-grained soil can withstand more water under high suction; therefore, the coefficients of permeability are usually smaller than coarse-grained soil. The value of hydraulic conductivity of soil changes according to soil type. The value of hydraulic conductivity depends on the size of the soil granule, the soil matrix structure, the type of fluid in the soil, and the degree of saturation of the soil. For soil matrix, it depends on the size distribution of the pore size, pore shape, and porosity, while for the fluid, it depends on the density and viscosity of the fluid.

The objective of this paper is to study the effect of soil hydraulic properties to the unsaturated soil behavior due to rainfall. Using the instrumentations, moisture

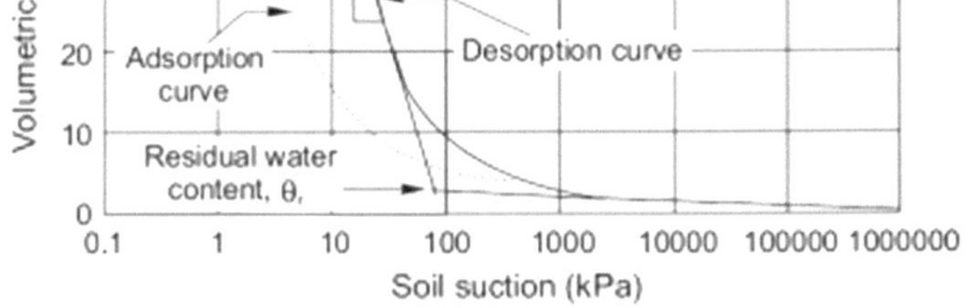

Fig. 1 Typical absorption and desorption of SWCC [3]

content will be attained. There were two tests in this experiment, and both results are presented and compared.

2 Materials and Methods

2.1 *Materials*

Two types of soil were tested in the experiment, sandy soil and kaolin, which were commercially obtained.

2.2 *Experiment Setup*

Undisturbed samples were used in the experiments. The experiment apparatuses include pressure plate extractor, constant head permeability, and soil column.

2.2.1 Basic Physical Properties

Basic soil properties, such as soil moisture content, bulk density, porosity, and water retention ability, were investigated in this research.

2.2.2 Pressure Plate Extractor Test (PPE)

The SWCC was determined in accordance with ASTM D6836-02 for determination of soil water characteristic curve for pressure extractor. The experiment is using 2 pressure chambers, a 5 bar, and a 10 bar with 20 bar air compressors and saturated ceramic plates. After samples were saturated for 24 h, the samples were placed on the saturated ceramic plate. The ceramic plate is attached to the water tube for water drainage while samples are on pressure (refer to Fig. 3). Then, lid of the chamber is closed and pressure is applied to the chamber and samples. After having last pressure applied to the samples, the samples are oven dried for 24 h and weighed again. In this test, the range of pressure applied to the samples is from 0 to 15 bar.

2.2.3 Constant Head Permeability

In this study, saturated hydraulic conductivity is determined by the constant head permeability method, in accordance with the procedures outlined by ASTM D2434-68 (2006). This method is chosen due to the vertical one-dimensional head that can be produced as soil column apparatus. The sample of soil in the mold is soaked into water to reach the saturation point for 24 h. The bottom of the sample is then wrapped with nylon cloth to prevent the movement of the soil from the mold during the experiment.

2.2.4 Infiltration Experiment

Experiment of infiltration was prepared by using soil column as a main apparatus and it was supported by other equipments, such as water supply and effluent system. 5 ft high and 8 in. diameter transparent acrylic cylinder soil column was used as shown in Fig. 2. In this experiment, volumetric water content and pore water pressure were measured using time-domain reflectometry (TDR) system and tensiometers, respectively. All collected data were logged to data acquisition system and retrieved

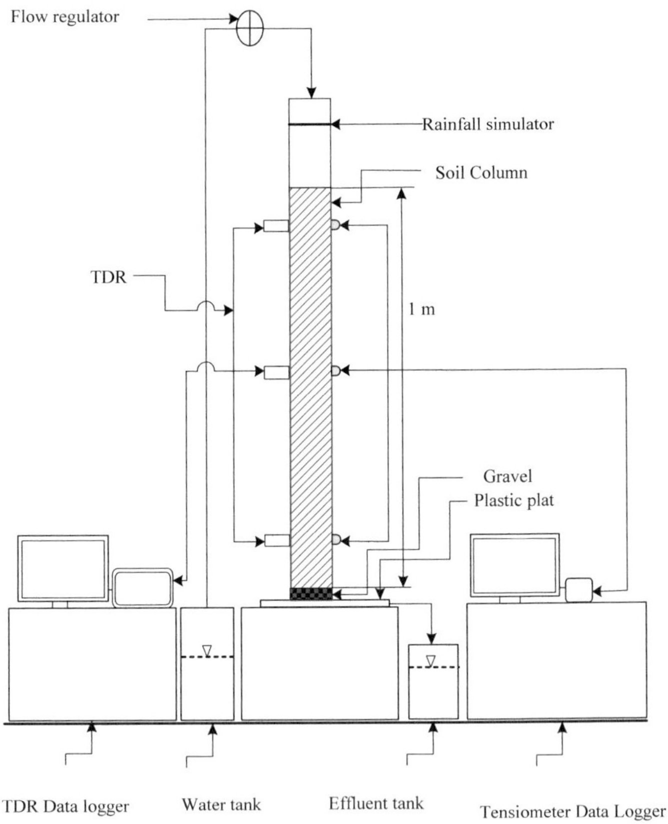

Fig. 2 Schematic diagram of soil column apparatus

Table 1 Experiment schedule for rainfall intensity and duration

No.	Sample no.	Rainfall duration (hours)	Rainfall intensity (mm/hr)
1	RS11,RK11	3	18.54
	RS12,RK12	6	
	RS13,RK13	12	
	RS14,RK14	24	
2	RS21,RK21	3	3.48
	RS22,RK22	6	
	RS23,RK23	12	
	RS24,RK24	24	

through personal computer attached to the loggers. In this study, infiltration tests were designed to meet the objectives mentioned earlier, which are shown in Table 1.

3 Results and Discussions

3.1 Soil Basic Properties

The results of the soil properties are summarized in Table 2.

3.2 Soil Hydraulic Properties

3.2.1 Hydraulic Conductivity (K)

Result obtained in Fig. 3 and Table 1 shows that hydraulic conductivity for sandy soil is higher than kaolin.

3.2.2 SWCC

Figure 3 (a) and (b) shows the result of SWCC presented by the curve obtained. It is indicated that decreasing the water for sand is more rapid than kaolin, which shows water retention ability of sand is lower than the kaolin. The parameters of volumetric saturated moisture content and volumetric residual moisture content are higher for kaolin than sandy soil (Table 1).

Table 2 Basic properties of the soils

Parameter	Unit	Sandy soil	Kaolin
Atterberg limit			
Plastic limit	%	–	31.9
Liquid limit	%	–	56
Plasticity index	%	–	24
Soil composition			
Gravel	%	0.1	0
Sand	%	97.5	4.3
Silt	%	2.4	40
Clay	%	0.1	55.7
Soil classification		SP	MH
		Poorly graded sand	Sandy elastic silt
Bulk density (ρ_b)	kg/m^3	1392	1106
Porosity (e)		0.47	0.58
Moisture content	%	0.10	0.37
Volumetric saturated moisture content, θ_s		0.07	0.48
Volumetric residual moisture content, θ_r Saturated hydraulic conductivity, K_{sat}	m/s	1.79×10^{-7} 1.14×10^{-3}	2×10^{-6} 2.1×10^{-6}

3.3 *Infiltration Result*

3.3.1 Sandy Sand

Result shows that moisture content value varies for all parts of columns as each time rainwater applied into soil column for intensity number 1. Wetting at the initial stage of soil column leads to the rapid increment of moisture content when RS11 is performed. The value is stable throughout the remaining wetting process.

Nevertheless, after RS11 was stopped, moisture content was progressively declined during the drying process. Moisture content of the soil will start increasing again for each time RS12, RS13, and RS14 are applied and it decreases during drying process. Result of intensity number 2 applied to the soil shows the same pattern as intensity number 1 for all samples. For location comparisons, the value at the top has significant changes followed by the center and bottom.

3.3.2 Kaolin

Figure 4 shows that the value increment for moisture content started from the top section, followed by the middle and lower section of column, respectively, for RK13, RK14, and RK23 rainfall. For intensity 1, the result distribution shows that after

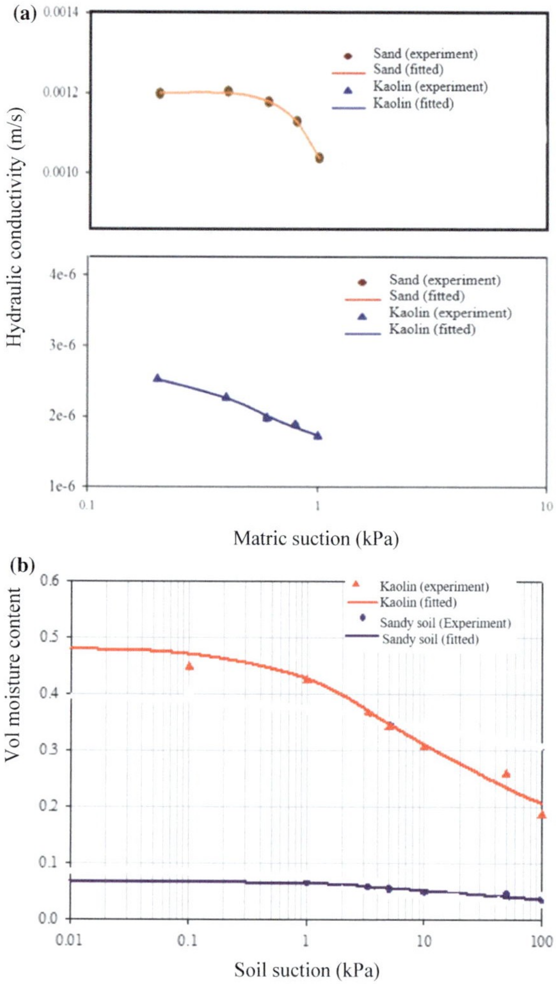

Fig. 3 Result of (**a**) Hydraulic conductivity, and (**b**) SWCC

RK12, the value of the top reaches the highest value. However, after RK13, the value in the middle continues to decline, followed by the upper and lower levels. Indefinable changes occurred for all sections of RK14 up to RK22. However, the value continues to experience the lowest value, followed by the bottom and the top. Change in value can be seen after RK24, and during drying process, the center value is the lowest followed by the bottom and the top. RK12 changed only after 12 h of rain on the top and middle of the soil column.

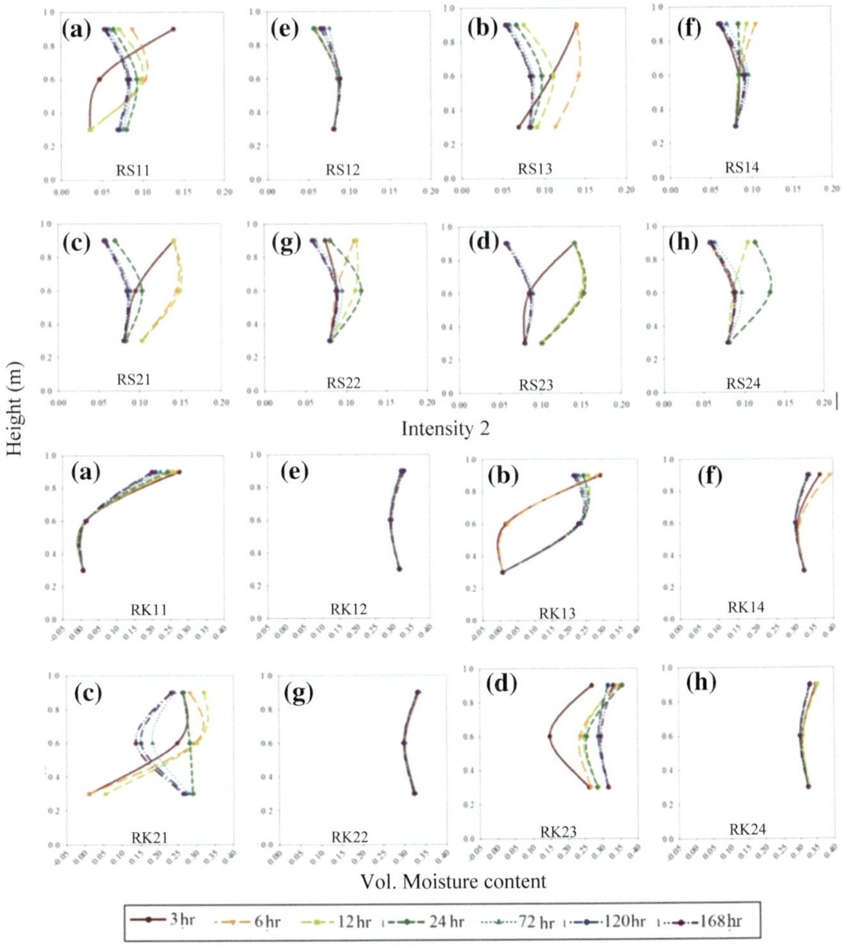

Fig. 4 Result of moisture content for silty sand and kaolin from infiltration test

4 Conclusion

For the SWCC results, the highest degree of saturation content obtained is kaolin, followed by sandy soil. These results show that kaolin has the ability to withstand water more than sand. Whereas for hydraulic conductivity values, kaolin soil has the lowest hydraulic conductivity value compared to sand soil. This indicates that the soil with smaller particles has a smaller water permeability value. Volume moisture content increases when water flow is applied to the soil and decreases during drying process. It was found that the middle part had the highest value of moisture content for both soils, while the value of the top had the lowest moisture content for both soils.

References

1. Agus SS, Leong EC, Rahardjo H (2001) Soil-water characteristic curve of Singapore residual soils. Geotech Geol Eng 19:285–309. Am J Environ Sci 4(4):303–309
2. Bujang BK, Ali FH, Low TH (2006) Water infiltration characteristics of unsaturated soil slope and its effect on suction and stability. Geotech Geol Eng 24:1293–1306
3. Fredlund DG, Xing A (1994) Equations for the soil-water characteristic curve. Can Geotech J 31:521–532
4. Gofar N, Lee LM, Kassim A (2008) Instrumented soil column model for rainfall infiltration study. In: GEOTROPIKA
5. Nam S, Gutierrez M, Diplas P, Petrie J, Wayllace A, Lu N, Munoz J (2009) Comparison of testing and models for establishing the SWCC of riverbank soils. Eng Geol 110:1–10
6. Pham K, Kim D, Choi H, Lee I, Choi H (2018) A numerical framework for infinite slope stability analysis under transient unsaturated seepage conditions. Eng Geol 243:36–49
7. Rahardjo H, Ong TH, Rezaur RB, Leong EC, Fredlund DG (2010) Response parameters for characterization of infiltration. Environ Earth Sci 60:1369–1380

Effect of Crumb Rubber Particles on the Properties of Asphalt

Ahmed Salama Eltwati, Amir Hossein and Danil Nasr

Abstract The practice of crumb rubber derived from scrap tires to modify the mechanical properties of bituminous mixtures has become increasingly important in road engineering. In recent years, much research has been devoted to the influence of this waste material on pavement performance. This paper aims to address the influence of crumb rubber (CR) polymers on the performance of asphalt pavements. Thus, asphalt mixtures with varying percentages of CR mixed by the wet process were tested. The statistical analysis revealed that crumb rubber significantly increased the stiffness modulus, rutting resistance, and pavement resistance to moisture damage. The study concludes that an addition of 20% to 24% of crumb rubber modifiers to conventional asphalt mixture yields the most satisfactory results among other percentages of CR polymers. The study recommends the use of crumb rubber in pavement construction.

Keywords Asphalt · Crumb rubber · Modified asphalt · Scrap tires · Particles

1 Introduction

The increase in the number of vehicles in recent times has led to increased road usage, resulting in higher volume of traffic and tire pressure on roads [1]. The resulting effect has premature deformation and failure in asphalt pavements [2, 3], generating serious concern among many highway and State agencies [4, 5]. To combat this challenge,

---

A. S. Eltwati (✉)
Department of Road and Airport Engineering, Faculty of Aeronautics, Bright Star University, El-Brega, Libya
e-mail: Ahmed.Eltwati@bsu.edu.ly

Department of Civil Engineering, Faculty of Engineering, University of Benghazi, Benghazi, Libya

A. Hossein
Department of Civil Engineering, Lenjan Branch, Islamic Azad University, Isfahan, Iran

D. Nasr
Department of Civil Engineering, Khorasgan Branch, Islamic Azad University, Isfahan, Iran

M. Awang and M. R. Meor M Fared (eds.), *ICACE 2019*,
Lecture Notes in Civil Engineering 59, https://doi.org/10.1007/978-981-15-1193-6_5

most governments have adopted the use of additives and modifiers in enhancing the performance of asphalt pavements [6, 7].

Due to the increasing trend of road usage by motorists, scrap tires have also been on the increase [8]. Disposing of these scrap tires poses a serious environmental concern as they are mostly burned or left at the mercy of nature to decompose as is prevalent in most developing countries [9]. Therefore, the need for an effective technique for recycling discarded tires has been considered by most countries [10]. One technique is to convert these discarded tires to crumb rubber (CR) [11]. CR, developed from discarded tires, is made up of natural or synthetic rubber [12]. The crumb rubber used in asphalt applications improves the performance of asphalt pavements and reduces traffic noise and improves resistance to cracking and rutting [13, 14].

The process of using crumb rubber to improve the performance of hot mix asphalt (HMA) is not a new practice and has been applied over the years [15]. Crumb rubber is mostly blended in the asphalt mixture by two methods, usually referred to as the wet and the dry process [16]. In the first method (wet process), the CR is mixed with bitumen at a temperature ranging from 191 to 218 °C for a period of about 1 to 2 h [17]. In this process, the bitumen is initially modified before it is included in the mixture, which in turn modifies the mixture property too [18]. On the other hand, in the dry process, the CR is first combined with aggregates and then mixed with bitumen [19]. The dry process requires around two to four times the amount of crumb rubber than that used in the wet process [20]. Among these two techniques, the wet method is the most widely used [21]. The distinguishing difference between the two methods is the binding property, i.e., the binding property is more effectively modified in the wet process than in the dry process, because of its direct interaction with the crumb rubber particles [18].

This paper investigates the influence of the CR content on the performance of asphalt pavements by using a wet process. It examines this effect by using an asphalt mix including different CR percentages on the behavior of asphalt surface, withstanding deformation and moisture damage by applying a wet method. The conventional 60/70 bitumen was utilized in the research mixed with several amounts of CR: 16, 20, 24, and 28% by weight of bitumen.

However, the resistance of rubberized and conventional asphalts to deformations (rutting) was assessed by applying the wheel-tracking experiment. Moreover, the moisture susceptibility test was adopted in this research in assessing the possible damages caused by moisture on asphalt mix. We employed the indirect tensile stiffness modulus (ITSM) test to ascertain the resistance of the asphalt mix to low-temperature cracking

The rest of the paper is organized as follows: Sect. 2 explains the materials and methodology employed in this research and in Sect. 3, the results of the experiments conducted were discussed. The topic is concluded in Sect. 4.

2 Materials and Methods

The materials used in this study were aggregates, bitumen, and crumb rubber.

2.1 *Aggregate*

Aggregates were collected by crushing rocks from a quarry. Coarse aggregates, fine aggregates, and fillers were obtained by doing a sieve analysis. Other tests, such as specific gravity and theoretical maximum density (ASTM D2041), were used as well to examine the aggregate properties. Ordinary Portland cement was used as a filler. Dense-graded aggregates of the maximum nominal size of 16 mm frequently used in the wearing course of asphalt pavement were implemented in the study [22].

2.2 *Bitumen*

The bitumen used was normal 60/70 grade.

2.3 *Crumb Rubber (CR)*

The CR used in this study was derived from waste tires with a particle size of 0.6–0.65 mm. Adding several percentages of CR: 16, 20, 24, and 28% by weight of bitumen to asphalt mixtures was used in this study.

2.4 *Asphalt Mixture Preparation*

The wet technique was used in this study. The CR was mixed with bitumen at a temperature of 195 °C for a period of 1 h before being added to the aggregates [23]. Mixtures of asphalt specimens of CR (16, 20, 24, and 28% by weight of bitumen) and conventional mixture (without additive) were set up for Marshall experiment. The asphalt samples that were made in Marshall molds had a width of 101.6 mm according to ASTM D.1559 [24]. The Marshall samples, each containing crumb rubbers, were then subjected to 75 compacting blows on each side of the specimen. The result of this test yielded the optimum bitumen content (OBC) for each sample.

3 Experimental Results and Discussion

Marshall stability test was carried out to evaluate the effect of crumb rubber contents on the maximum load carried by a compacted specimen at a temperature of 60 °C. The wheel-tracking test was conducted to ascertain the performance of various rubberized asphalt mixtures under loading, as well as their resistance to rutting. The effect of moisture on rubberized asphalt mixtures was evaluated by using the moisture susceptibility test. The indirect tensile strength test was used to evaluate the effect of crumb rubber on the bearing capacity of mixtures.

3.1 Marshall Test

The optimum bitumen content (OBC) for all mixtures was obtained by using Marshall test (ASTM D1559-89). The results indicated that an unmodified asphalt mixture (AC 60–70) has the lowest OBC (5.04% of the mixture weight). On the other hand, mixtures with 24% and 28% of CR polymers have the OBC of 6.1% and 6.4%, respectively (values are slightly higher due to the addition of rubber particles). Figure 1 shows the bitumen contents for all rubber-modified asphalt mixtures.

Figure 2 displays the effect of CR addition on the mixture stability. It shows that stability slightly increases as CR contents in the mixture increase from the findings, the stability of asphalt mix is largely affected by the interior friction of aggregates and the viscosity of bitumen, and hence the viscosity of bitumen increases because of adding crumb rubber to the bitumen.

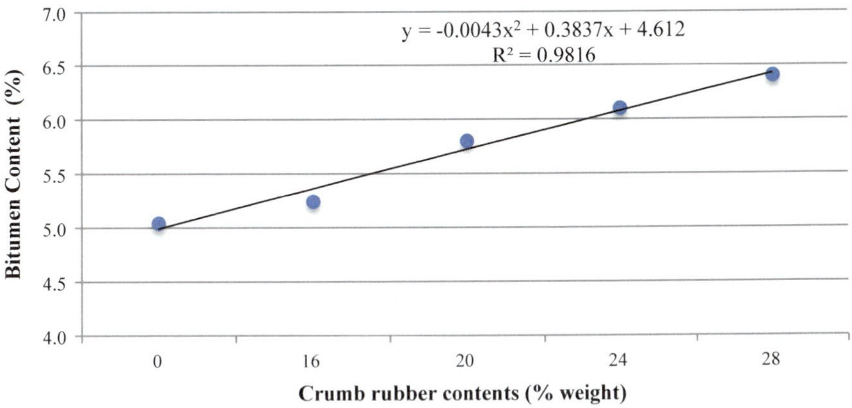

Fig. 1 Effect of CR on bitumen content in asphalt mixtures

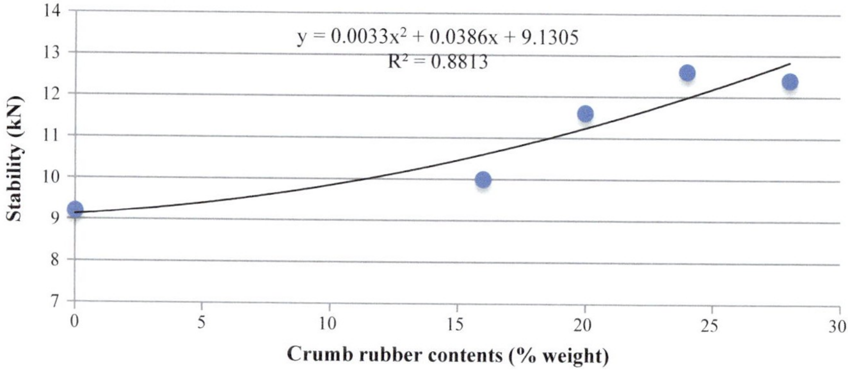

Fig. 2 Effect of CR contents on marshall stability

Table 1 presents the statistical analysis of T-test analysis ($\alpha = 0.05$) of stability test results of different crumb rubber percentages. The analysis shows that P-value is larger than 0.05 and t-stat. is smaller to t-Critical, which means CR has no significant effect on the stability of asphalt binder.

Although a previous study [21] asserts to the result of this study, i.e., that an addition of crumb rubber enhances the properties of hot asphalt mixture and increases the pavement stability by using a dry process, it, however, failed to show in detail the effect of various values of crumb rubber on properties of hot asphalt mix. In addition, in another study [16], the added CR to bitumen by using dry process shows that the Marshall stability of the sample decreases as CR increases. This study found that the CR particles are not totally liquefied in the bitumen, and perform elastically inside the mixture. This result contradicts with this study's result. We, therefore, conclude that using different mix processes, dry and wet, produce different stabilities when CR is added to bitumen.

3.2 *Wheel-Tracking Test*

The wheel-tracking test assesses the pavement resistance to rutting (deformation) by the repeated passing of a loaded wheel. Once the loaded wheel begins moving over the pavement model, the pavement starts resisting the rutting. However, after a certain number of repetitions of the loaded wheel, the rutting has occurred. By this

Table 1 Output from statistical analysis of T-test on influence of crumb rubber content on stability

	CR %	Stability (kn)
t-Stat.	1.319805	
P(T <= t)	0.128683	
t-Critical	2.131847	

method, we examined the effect of utilizing different amounts of CR in a wet process mixed by assessing the obtained results with that of conventional mixtures.

The samples of different CR contents were compacted in standard molds having dimensions of 410 × 255 mm and height of 55 mm. Each sample was made with a minimum density of 98%. Then, the compacted specimens were kept in an oven at a temperature of 60 °C for about 4 h. The loaded wheel was adjusted to a velocity of 40 cycles per minute. The experiment was performed for 10,000 cycles. The rutting (deformation) depth in the pavement model was recorded at certain cycles as depicted in Fig. 3.

Figure 3 illustrates that as the amount of crumb rubber increases, the rutting depth diminishes, which verifies that the mixing asphalt with CR will enhance the pavement performance. Figure 4 reveals that deformation depth reached to lowest value at 24% of CR and then increases again. Therefore, the conventional asphalt mixture AC16 with crumb rubber of 24% yields the highest rutting resistance in asphalt pavement.

The statistical analysis shown in Table 2 states that P-value is smaller than 0.05 and t-stat is larger than t-Critical, which indicates that adding various percentages of CR to asphalt binder significantly boosted the pavement resistance to the rutting.

This result agrees with previous research conducted by using the wheel-tracking test [25], where the modified asphalt mixes were observed to be more resistant to the rutting, but it has not shown in detail the effect of using different crumb rubber contents.

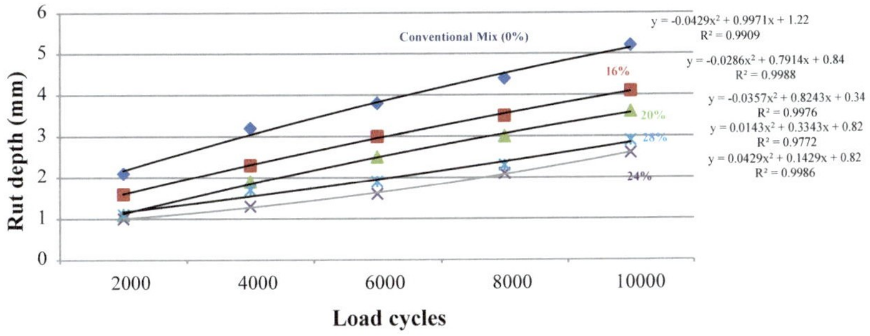

Fig. 3 Rutting in asphalt pavements by using the wheel-tracking test

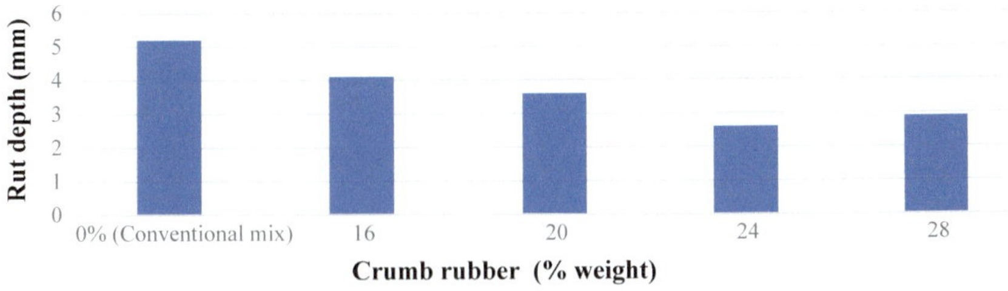

Fig. 4 Effect of addition of CR on pavement resistance to rutting (after 10,000 cycles)

Table 2 Output from T-test analysis on influence of CR content on pavement resistance to rutting

	CR %	Rut depth (mm)
t-Stat	2.867	
P(T <=t)	0.023	
t-Critical	2.132	

3.3 *Effect of Moisture on Asphalt Mix*

The moisture susceptibility test was done in this study to examine the possible moisture damage in asphalt pavement and to reveal the effect of using different percentages of CR. This test was conducted in accordance to ASTM D4867M. The finding of this assessment presents the tensile strength ratio (TSR) for different samples as illustrated in Fig. 5.

Figure 5 clearly reveals that asphalt mixtures containing CR have greater values of TSR than conventional asphalt mixtures (0% CR), which indicates that the modified asphalt mixtures have better resistance to moisture damage. The results show that bitumen mixed with CR of 20% produced the highest value of TSR, and thus yields the best resistance to moisture damage.

The statistical analysis in Table 3 states that P-value is much smaller than 0.05 and t-stat. is larger than t-Critical, which proves that addition of CR to asphalt binder significantly improved the pavement resistance to moisture damage.

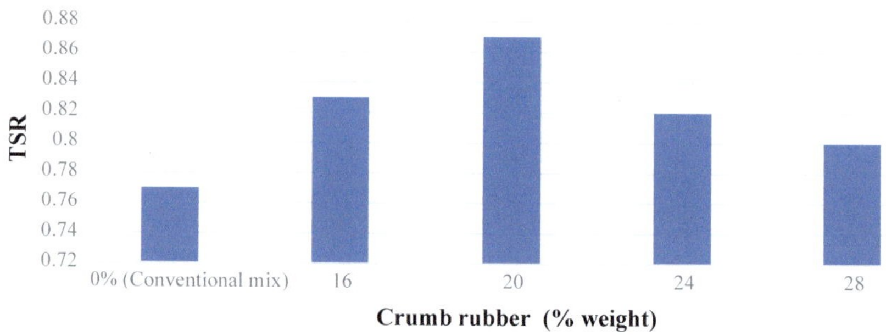

Fig. 5 Result of the moisture susceptibility test

Table 3 Output from T-test analysis on influence of crumb rubber content on moisture damage

	CR %	TSR
t-Stat	3.472	
P(T <=t)	0.013	
t-Critical	2.132	

The result of this study is consistent with a past study [26] in which an addition of crumb rubber to unmodified mixture increases the resistance to moisture damage, but the optimum CR percentages were not consistent. In this study, crumb rubber of 20% by weight of bitumen gave the best performance, while in the previous study [26], CR of 10% provided the best result.

3.4 Indirect Tensile Strength of Asphalt Mixtures

This test was used to estimate the potential occurrence of rutting and evaluate how the bearing capacity of the mixtures is influenced by the addition of crumb rubber. The test was done in accordance with standard specification ASTM D6931–12. The test mold, which had a diameter of 101.6 mm, was manufactured with an impact compactor at a temperature of 25 °C. After pouring asphalt mixture into the test mold, the mixture was applied to loading with a rate of 50 mm/min. The rutting in the mixture is recorded as this loading increases. Figure 6 shows the result of this test.

Figure 6 reveals that mixtures with crumb rubber addition had larger stiffness modulus than those without any CR. This is an indication that crumb rubber-modified mixtures are more resistant to the rutting than conventional asphalt binders. Furthermore, adding a percentage of 24 of crumb rubber to the mixture produced the best result.

Table 4 displays the statistical analysis of T-test of indirect tensile strength test results. The table shows that P-value is considerably smaller than 0.05 and t-Stat is larger than t-Critical, which means that addition of CR to asphalt binder significantly raised the stiffness modulus and the pavement is thus less affected by traffic loads.

The result of this study is consistent with past studies [27] where CR was added to a mixture by the dry process. We, therefore, conclude that the effect of CR on the mixture stiffness modulus is not affected by process methods: dry or wet.

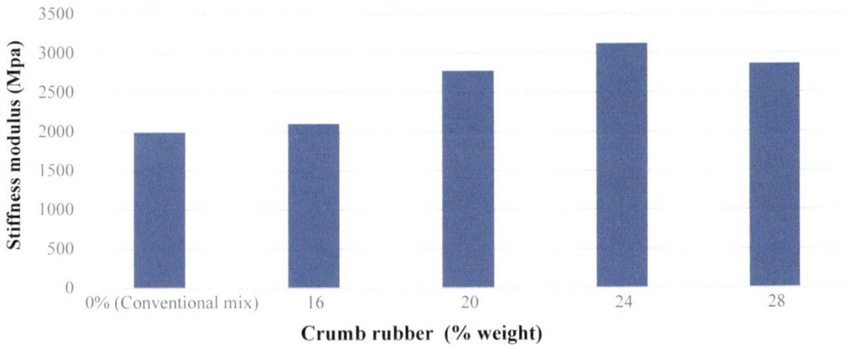

Fig. 6 Result of indirect tensile stiffness modulus test

Table 4 Output from T-test analysis on influence of CR on stiffness modulus

	CR %	Stiffness modulus (Mpa)
t-Stat	−11.493	
P(T <=t)	0.0002	
t-Critical	2.132	

4 Conclusion

In this paper, the effect of using different crumb rubber percentages in the asphalt mixture by wet process technique is presented. The study involved utilizing different rubber percentages (16, 20, 24, and 28%) by weight of bitumen. A comparative assessment was made to evaluate these different crumb rubber percentages to the conventional asphalt mixture. Based on the findings of this study, we conclude that

- The statistical result exhibited that an addition of crumb rubber to bituminous mixtures significantly improved the asphalt pavement resistant to rutting. Asphalt mixture modified with 24% CR yielded the highest resistance to rutting.
- It was shown that adding crumb rubber to bituminous mixtures significantly increased the tensile strength ratio (TSR), and thus it is more resistant to moisture damage. Mixture modified with 20% CR provided the highest value of TSR, and thus it is less prone to moisture damage.
- The experimental result indicated that the impact of adding of CR on the pavement resistance to the moisture damage was not affected by the process methods: dry or wet.
- The statistical analysis showed that adding CR to asphalt mixtures significantly enhanced the stiffness modulus. Therefore, the pavement has a much larger bearing capacity when subjected to heavy traffic loads.
- Adding crumb rubber to bituminous mixes not only reduces the accumulation of scrap tires at landfills, but it may also be an effective means of improving the long-term performance of pavements.

References

1. Ben-Edigbe JA, Fallah I, Minhans A (2011) Influence of composite traffic control mechanisms on four-arm highway intersection. In: Proceedings, annual conference—Canadian society for civil engineering. Canada's University (uOttawa)
2. Abdelgalil A, Nor H (2014) The influence of block thickness on the performance of asphalt paving blocks. Res J Appl Sci Eng Technol 7(10):2129–2134
3. Irfan M et al (2018) Performance evaluation of crumb rubber-modified asphalt mixtures based on laboratory and field investigations. Arab J Sci Eng 43(4):1795–1806
4. Croney P, Croney D (1998) Design and performance of road pavements, 3rd edn. McGraw-Hill, New York, 231 p
5. Ferrotti G, et al (2017) Effect of warm mix asphalt chemical additives on the mechanical performance of asphalt binders. Mater Struct 50(5):226

6. Moon KH et al (2014) Using recycled asphalt materials as an alternative material source in asphalt pavements. KSCE J Civ Eng 18(1):149–159
7. Li H, et al (2019) Physical, rheological and stability properties of desulfurized rubber asphalt and crumb rubber asphalt. Arab J Sci Eng
8. Ben-Edigbe J et al (2012) Extent of delay and level of service at signalised roundabout. Int J Eng Technol 2(3):419–424
9. Wang W et al (2018) Laboratory investigation on high- and low-temperature performances of asphalt mastics modified by waste oil shale ash. J Mater Cycles Waste Manag 20(3):1710–1723
10. Vacková P, Valentin J, Hamzah MO (2019) Asphalt concrete performance combining use of increased reclaimed asphalt content and warm mix additives. Springer International Publishing, Cham
11. Abdelhak B et al (2016) Effect of recycled asphalt aggregates on the rutting of bituminous concrete in the presence of additive. Arab J Sci Eng 41(10):4139–4145
12. Heitzman M (1992) Design and construction of asphalt paving materials with crumb rubber modifier. Transp Res Rec 1339:1–8
13. Chandran V, et al (2015) Evaluation of performance of natural rubber composites with different sizes of waste tyre rubber (WTR) and precipitated silica on C–M–M. Arab J Sci Eng 40(4):1187–1196
14. Abdelgalil A, H.J.R.J.o.A.S. Nor, Engineering, and Technology (2014) The influence of block thickness on the performance of asphalt paving blocks. J Appl Sci Eng Technol 7(10):2129–2134
15. Putman BJ, Amirkhanian SN (2006) Crumb rubber modification of binders: interaction and particle effects. In: Proceedings of the asphalt rubber 2006 conference
16. Hernández-Olivares F et al (2009) Rubber-modified hot-mix asphalt pavement by dry process. Int J Pavement Eng 10(4):277–288
17. Takallou H, McQuillen J, Hicks R (1985) Effect of mix ingredients on performance of rubber-modified asphalt mixtures. Alaska DOT and Public Facilities, Fairbanks, AK
18. Airey GD, Rahman MM, Collop AC (2003) Absorption of bitumen into crumb rubber using the basket drainage method. Int J Pavement Eng 4(2):105–119
19. Wong CC, Wong W-G (2007) Effect of crumb rubber modifiers on high temperature susceptibility of wearing course mixtures. Construct Build Mater 21(8):1741–1745
20. Roberts FL, et al (1989) Investigation and evaluation of ground tire rubber in hot mix asphalt. National Center for Asphalt Technology
21. Moreno F, Rubio M, Martinez-Echevarria M (2011) Analysis of digestion time and the crumb rubber percentage in dry-process crumb rubber modified hot bituminous mixes. Construct Build Mater 25(5):2323–2334
22. Asphalt Institute (1988) The asphalt handbook. 1989 edn. Manual series, vol xxxi. Asphalt Institute, College Park, Md., USA, 608 p
23. Buncher MS (1995) Evaluating the effects of the wet and dry processes for including crumb rubber modifier in hot mix asphalt. DTIC Document
24. ASTM (2001) Road and paving materials; vehicle-pavement systems. American Society of Testing Materials, West Conshohocken, Pennsylvania
25. Fontes LPTL et al (2010) Evaluating permanent deformation in asphalt rubber mixtures. Construct Build Mater 24(7):1193–1200
26. Qiu X, Chen LY, Xue L (2011) Effect of crumb rubber modifier on pavement performance of wearing course asphalt mixture. Adv Mater Res 168:1145–1148
27. Moreno Navarro F, Rubio Gamez MC (2012) Influence of crumb rubber on the indirect tensile strength and stiffness modulus of hot bituminous mixes. J Mater Civ Eng 24(6):715–724

Issues and Adoption Process for Supply Chain Management of Materials in Construction Projects

Sulaiman Hasim, Mohd Azrizal Fauzi, Intan Rohani Endut, Zulkiflee Yusof and Ahmad Ruslan Mohd Ridzuan

Abstract The supply chain (SC) is a new term that emphasizes the interaction between marketing, logistics, and production. With the adoption of the SC comes the opportunity primarily related to the management of procurement of logistics materials across corporate borders, such as between firms and their suppliers. This paper presents the issues and the lifting process for supply chain management in construction projects. Discussions on issues faced by contractors relating to material procurement process have also been captured and enhancement strategies related to management, human relations, and information and communication technology have been brought forward to improve the performance of SCM materials. This study suggests that the management of material supply chains can be enhanced through management integration, human relations, and information and communication technology. This paper also considers the potential of implementing SCM to integrate the construction process in Malaysia, and therefore addressing important issues including the vague definition of what is needed from the client, lack of communication between parties involved in the material supply chain, ambiguity between plans and quantity bills, and low cost associated with the traditional process of material procurement.

Keywords Supply chain management · Construction projects · Contractors · Technology · Human Relationship

S. Hasim (✉) · M. A. Fauzi · I. R. Endut · Z. Yusof · A. R. M. Ridzuan
Faculty of Civil Engineering, Universiti Teknologi MARA, 40450 Shah Alam, Selangor, Malaysia
e-mail: sulaimanhasim@yahoo.com

M. A. Fauzi · I. R. Endut · Z. Yusof · A. R. M. Ridzuan
Faculty of Civil Engineering, Universiti Teknologi MARA, Cawangan Pulau Pinang, 13500, Permatang Pauh, Penang, Malaysia

Institute for Infrastructure Engineering and Sustainable Management (IIESM), Universiti Teknologi MARA, 40450 Shah Alam, Selangor, Malaysia

M. Awang and M. R. Meor M Fared (eds.), *ICACE 2019*,
Lecture Notes in Civil Engineering 59, https://doi.org/10.1007/978-981-15-1193-6_6

1 Introduction

For increased productivity in construction projects, material management is an important function. Material planning and eradication, vendor evaluation and selection, purchase, expense, shipping, material acceptance, warehousing and inventory, and material distribution are the key material management functions [1]. Project cost, time, and quality are influenced by improper handling and material handling during the construction process [2]. Furthermore, the total construction cost ranges from 30 to 80% of the cost for material handling [3] and almost 60% of the total working capital of any industrial organization consists of material costs [4]. Thus, to control productivity and costs in construction projects efficient material management is required. In construction projects, there are many issues that contribute to poor material management. Some of the adverse effects on material management are with excessive working papers, misuse of specs, improper handling of the site, waste, and lack of proper work plans [5].

This paper reviews the issue of current material management practices in construction projects and explores past research in management, human relations, and information and communications technology being implemented. It begins with a review of material management practices in construction projects and common problems associated with them. The way wherein the materials are managed in the project is deliberated and the stretches for improvements are emphasized.

2 Issues Encountered by Contractors

Based on preliminary interviews and a literature review, fifteen (15) issues on material supply chain management have been identified. Among these issues are the fuzzy definition required by the client [6]; lack of communication between parties involved in the material supply chain [7]; uncertainty between quantity and quantity bill [8]; having too many materials suppliers [9]; the absence of building materials required [10]; delayed submission of building materials approval by contractor to consultants [11, 12]; submitting on-site building materials storage of work [13, 14]; limited building material storage on site [15]; loss of building materials delivered to the site [16]; theft of materials on the site [17]; vandalism at the site [18]; damage during handling of materials (defects); weak communication between parties involved in the supply chain at the site [19]; and accept, operate, and store unused building materials.

SCM aims to create the maximum value, not just for the company, but also for the entire network of the supply chains including the end customers. Hence, to improve the efficiency and effectiveness of the process for the entire supply chain integration, supply chain and recycling process should be designed. It is important that benefits be distributed fairly. References [20–22] advised the government to resolve these issues before the country could realize its intention to become a logistics hub in the region.

3 Adoption of Material Supply Chain

This section discusses the improvement strategies in adoption of material supply chain using traditional contract in a construction project. Three (3) areas were identified by the researcher through literature review, which can be classified as management, human relationships, and information and communication technologies.

3.1 Management

Based on the previous studies these are among the best strategic management aspects that have been adopted:

- Implementing a project with own resources [23].
- Identify areas of improvement in material supply chains by top management [24].
- Commitment from top management to improve process supply chain materials [25].
- Strategic network optimization including numbers, locations, and facilities in the material supply chain [26].
- Coordinate overall organizational strategy with material supply chain processes [25].
- Contractors and suppliers need to be involved in the design team at the early stage [27].
- Managing process management of the material supply chain correctly [28].
- Suitable acquisition path-effective delivery [29].
- Lead times, cost reduction, and profit engagement [30].
- Decline in material prices and global currency issues [31–33].
- Political involvement in controlling material costs [32, 34].

3.2 Human Relationships

Supply chain relationships play a significant role in achieving the firm's goals. The coordination and integration of activities with suppliers and understanding of customer's needs result in greater benefits for companies and the aspect adopted as below:

- Long-term relationship—trust, experience, and responsiveness [35].
- Reward creativity, encouraging development, and implementation of new ideas [36].
- Resources development in materials supply chain [37].
- Knowledge asset in materials supply chain [38].

- Long term and needs resources commitment [39].
- Conduct training for suppliers [40].
- Rationalization of existing suppliers [41].
- Set up review mechanisms for feedback, ideas, and knowledge exchange [42].
- Potentially beneficial effects of long-term trust value [43].
- Strategic partnership with distributors [44].
- Enhance a partnership with existing suppliers [45].
- Active participation of suppliers during construction phase [46].
- Encourage teamwork with the supply chain partners [47].
- Encourage problem-solving with the supply partners [47].

3.3 Information and Communication Technology

Telecommunication and computer technology enable all construction project players in the supply chain to communicate with each other. The use of information technology enables suppliers, manufacturers, distributors, retailers, and customers to reduce lead time, paperwork, and other unnecessary activities. Below are the key benefits of adopting information technology and communications for construction projects:

- Information technology chain operation system enables sharing of information [48].
- Ensure procedures for communication effectively [49].
- Firms' cultures toward innovation—implementation of new ideas of materials supply chain processes effectively [50].
- Establish the system for project participant to communicate effectively [51].
- Technology-oriented material suppliers [52].
- Establish database where to make and make-buy decision [53–55].
- Centralized information system [56].
- Establish radio frequency identification (RFID) for material tracking [57, 58].
- Implement building information modeling (BIM) to enhance projects [59–61].
- Implement geographic information system (GIS) for each project [62].

4 Conclusions

Since the early 1960s, the concept of SCM has been tied to distribution and physical logistics to illustrate the management of product flow activities. What's new is the emphasis given to border management. The material handling function in the construction industry is often fragmented with minimum communication and no clear responsibilities among the parties involved. The difficulty is still in reaching the expected benefits. Because cooperation in traditional methods is usually among members with different or different reward systems, experts need to realize the benefits of their cooperation. Balancing this benefit should be in such a way so that all members better in their cooperation receive more benefits, which can make a new challenge for SC managers. The improvement strategies involving such as management, human relationship, and ICT would increase productivity, reduce cost, and improve efficiency of construction projects.

References

1. Bell LC, Stukhart G (1986) Attributes of materials management systems. J Constr Eng Manag 112(1):14–21
2. Putra CW, Ahmad A, Majid MZA, Kasim N (1999) Improving material scheduling for construction industry in Malaysia. Malays Sci Technol Congr 99:6–8
3. Proverbs DG, Holt GD, Olomolaiye PO (1999) Construction resource/method factors influencing productivity for high rise concrete construction. Constr Manag Econ 17(5):577–587
4. Dey AK (2001) Understanding and using context. Pers Ubiquitous Comput 5(1):4–7. https://doi.org/10.1007/s007790170019
5. Zakeri M, Olomolaiye PO, Holt GD, Harris FC (1996) A survey of constraints on Iranian construction operatives' productivity. Constr Manag Econ 14(5):417–426
6. Pennanen A, Koskela L (2005) Necessary and unnecessary complexity. In: First international conference on complexity, science and the built environment, no. September, pp 1–12
7. Jaffar N, Tharim AHA, Shuib MN (2011) Factors of conflict in construction industry: a literature review. Procedia Eng 20:193–202
8. Yusuf GA, Mohamed SF (2016) Examination of pre-tender budgeting techniques for mechanical and electrical services in Malaysia. In: Proceedings of world academy of science, engineering and technology, no. 63
9. Sanayei A, Farid Mousavi S, Abdi MR, Mohaghar A (2008) An integrated group decision-making process for supplier selection and order allocation using multi-attribute utility theory and linear programming. J Franklin Inst 345(7):731–747
10. Ala-Risku T, Kärkkäinen M (2006) Material delivery problems in construction projects: a possible solution. Int J Prod Econ 104(1):19–29
11. Hamzah N, Khoiry MA, Arshad I, Tawil NM, Che Ani AI (2011) Cause of construction delay—theoretical framework. Procedia Eng 20:490–495
12. Marzouk MM, El-Rasas TI (2014) Analyzing delay causes in Egyptian construction projects. J Adv Res 5(1):49–55
13. Darvik L, Larsson J (2010) The impact of material delivery-deviations on costs and performance in construction projects
14. Okoye PU, Ngwu C, Ugochukwu SC (2015) Evaluation of management challenges facing construction practice in Nigeria. Int J Appl Innov Eng Manag 4(1):19–28
15. Khalil E, Abd-Elmohsen M, Anwar AM (2015) Impact resistance of rubberized self-compacting concrete. Water Sci 29(1):45–53
16. Sakurai Y, Mayhew P, White M (2008) Theft and vandalism at residential building sites in Australia
17. Farinloye I (2006) Disasters: man's plea of guilt. In: The punch, p 15
18. Farinloye O, Odusami K, Adewunmi Y (2013) Theft and vandalism control measures on building sites in Lagos, Nigeria. J Eng Proj Prod Manag 3(1)
19. Donyavi S, Flanagan P (2009) The impact of effective material management on construction site performance for small and medium sized construction enterprises. In: Proceedings of the 25th annual ARCOM conference. arcom.ac.uk

20. Lambert DM, Cooper MC (2000) Issues in supply chain management. Ind Mark Manag 29(1):65–83
21. Malaysian Productivity Centre (MPC) (2008) The Malaysian logistics industry as a crucial component of national competitiveness. In: Seminar on Enhancing the Competitiveness of the Logistics Industry. https://www.mima.gov.my/mima/wpcontent/uploads/MPC%20JB%20(3%20Jul08)%20Seminar%20on%20Enhancing%20%20the%20Competitivss%20of%20the%20Logistics%20Industry.pdf
22. Thong YC (2007) Overview-the logistics industry in the northern region and the need to enhance its effectiveness and competitiveness
23. Gholipour Y, V. T. Construction (2006) Multi-project resources procurement in the construction industry'. researchgate.net
24. Li S, Rao S, Ragu-Nathan TS, Ragu-Nathan B (2005) Development and validation of a measurement instrument for studying supply chain management practices. J Oper Manag. Elsevier
25. Wang X, Love PED, Kim MJ, Park CS, Sing CP, Hou L (2013) A conceptual framework for integrating building information modeling with augmented reality. Autom Constr 34:37–44
26. Melo MT, Nickel S, Saldanha-da-Gama F (2009) Facility location and supply chain management—a review. Eur J Oper Res 196(2):401–412
27. Song L, Mohamed Y, AbouRizk SM (2008) Early contractor involvement in design and its impact on construction schedule performance. J Manag Eng 25(1):12–20
28. Lambert DM, Croxton KL (2005) An evaluation of process-oriented supply chain management frameworks. J Bus Logist 26(1):25–51
29. Ojo A, Gbadebo M (2012) Critical selection criteria for appropriate procurement strategy for project delivery in Nigeria. J Emerg Trends Econ. journals.co.za
30. Paulsen C, Gallivan F, Chavez M, Venner M (2008) NCHRP 8-36 task 72: guidelines for cost estimation improvements at state DOTs
31. Polhemus N (2016) A Culture of commodification: hemispheric and intercolonial migrations in the trans-Atlantic Slave Trade, 1660–1807
32. Adeyeye S, Oyewole O et al (2015) A survey on traditional fish smoking in Lagos State, Nigeria. A J food. researchgate.net
33. Hussin J, Rahman I, Memon A (2013) The way forward in sustainable construction: issues and challenges. Int J Adv Appl Sci. researchgate.net
34. Elzarka S, Tipi N, Hubbard N et al (2008) Creating a logistics competency framework for Egyptian clothing companies. In: Logistics research network LRN. papers.ssrn.com
35. Naslund D, Williamson (2010) What is management in supply chain management?-a critical review of definitions, frameworks and terminology. J Manag Policy Pract. na-businesspress.com
36. Verschuren D et al (2002) History and timing of human impact on Lake Victoria, East Africa. Proc R Soc London Ser B Biol Sci 269(1488):289–294
37. Fisher K, Geenen J, Jurcevic M, McClintock K, Davis G (2009) Applying asset-based community development as a strategy for CSR: a Canadian perspective on a win-win for stakeholders and SMEs. Bus Ethics A Eur Rev 18(1):66–82
38. Rahman AA, Tay MY, Aziz YA (2016) Potential of knowledge management as antecedence of sustainable supply chain management practices. Int J Supply Chain Manag 5(2):43–50
39. Stone DL, Deadrick DL (2015) Challenges and opportunities affecting the future of human resource management. Hum Resour Manag Rev 25(2):139–145
40. Practices SB, Diversity S (2014) AT & T principles of conduct for suppliers, pp 1–4
41. AG S (2008) Code of conduct for Siemens suppliers declaration of the supplier. München
42. Cabrera EF, Cabrera A (2005) Fostering knowledge sharing through people management practices. Int J Hum Resour Manag 16(5):720–735
43. Coles C (2010) Trust and value through long-term market relationships: better than a short-term focus on price—Article—ODI Opinions 146 published by ODI, no. August
44. Ng T, Chung W (2008) The roles of distributor in the supply chain: push-pull boundary. Int J Bus Manag. pdfs.semanticscholar.org

45. Croxton KL, García-Dastugue SJ, Lambert DM, Rogers DS (2001) The supply chain management processes. Int J Logist Manag 12(2):13–36
46. Roth B (2012) Beckett. The reverse side of words. Topique 120(3):101–111
47. Gunasekaran A, Lai K, Omega TC (2008) Responsive supply chain: a competitive strategy in a networked economy. Elsevier
48. Hiltunen V, Kangas J, Pykäläinen J (2008) Voting methods in strategic forest planning—experiences from Metsähallitus. For Policy Econ 10(3):117–127
49. Rajhans RK (2013) Financial determinants of firm's value: evidence from Indian firms. ZENITH Int J Bus Econ Manag Res. papers.ssrn.com
50. Szczepańska-Woszczyna K (2014) Innovation processes in the social space of the organization Reg Form Dev Stud. 193.219.76.12
51. Baiden B, Price A, Dainty A (2006) The extent of team integration within construction projects. Int J Proj Manag. Elsevier
52. Nair SD, Ferron RD (2014) Set-on-demand concrete. Cem Concr Res 57:13–27
53. Bajec P, Jakomin I (2010) A make-or-buy decision process for outsourcing. PROMET Traffic Transp. hrcak.srce.hr
54. Davis G (2006) Management information systems (The Blackwell Encyclopedia of Management)
55. Benton T (2001) Philosophy of social science: the philosophical foundations of social thought
56. Sulankivi K (2004) "Benefits of centralized digital information management in multipartner projects. J Inf Technol Constr (ITcon). Citeseer
57. Wang X, Love P, Kim M, Park C et al (2013) A conceptual framework for integrating building information modeling with augmented reality. Elsevier
58. Lu M, Chen W, Shen X, Lam HC, Liu J (2007) Positioning and tracking construction vehicles in highly dense urban areas and building construction sites. Autom Constr 16(5):647–656
59. Arayici Y, Egbu C, Coates S (2012) Building information modelling (BIM) implementation and remote construction projects: issues, challenges, and critiques. J Inf Technol Constr. usir.salford.ac.uk
60. Azhar S, Khalfan M, Maqsood (2012) Building information modelling (BIM): now and beyond. Constr Econ Build. epress.lib.uts.edu.au
61. Wu S, Wood G, Ginige K, Jong S (2014) A technical review of BIM based cost estimating in UK quantity surveying practice, standards and tools. J Inf Technol Constr 19:534–562
62. Palve S (2013) Applications of GIS in infrastructure project management. J Struct Civ Eng Res. ijscer.com

Cement Hydration Extents for Hardened Cement Paste Incorporating Nanosized-Palm Oil Fuel Ash: A Thermal and XRD Analysis Study

Mohd Azrul Abdul Rajak, Zaiton Abdul Majid and Mohammad Ismail

Abstract The suitability of nanosized-POFA (nPOFA) as a supplementary cementing material has not been discovered enough. Therefore, the present study investigates the degree of cement hydration of Hardened Cement Paste (HCP) incorporating nPOFA via thermal and XRD analysis. The nPOFA was prepared by grinding the micro-POFA (mPOFA) using ceramic ball milling for 30 h. A series of Hardened Cement Paste (HCP) incorporating 10–60% of nPOFA and 10–30% of mPOFA were prepared and examined at 28- and 90-day curing age. The hydration of cement was assessed by monitoring the dehydroxylation of portlandite (CH) through Thermogravimetric Analysis (TGA) and X-ray Diffraction (XRD) analysis. The prepared nPOFA possesses average crystallite size calculated from the XRD diffractogram which is 34.74 nm, and BET surface area of 145.35 m^2/g. Based on the TG analysis, the HCP with 30% nPOFA replacement gives the highest CH consumption in the HCP within the interval time of two curing ages. Significantly, it indicates that 30% nPOFA replacement in HCP accelerates the hydration of cement through the pozzolanic reaction between the amorphous silica and CH from the hydrated products. Meanwhile, the XRD analysis indicated a decrease in CH peaks intensity in nPOFA HCP, which corresponds to the consumption of CH in the pozzolanic reaction. The results show that the high fineness of nPOFA results in the high distribution in HCP and rapid dissolution as it increases the rate of pozzolanic reaction.

Keywords nPOFA · Hardened cement paste · Pozzolanic reaction · Palm oil fuel ash

---

M. A. A. Rajak (✉)
Pusat Persediaan Sains & Teknologi, Universiti Malaysia Sabah, Jln. UMS, 88400 Kota Kinabalu, Sabah, Malaysia
e-mail: azrulrajak88@ums.edu.my

Z. A. Majid
Department of Chemistry, Faculty of Science, UTM, Johor Bahru, Johor, Malaysia

M. Ismail
Faculty of Civil Engineering, UTM, Johor Bahru, Johor, Malaysia

M. Awang and M. R. Meor M Fared (eds.), *ICACE 2019*,
Lecture Notes in Civil Engineering 59, https://doi.org/10.1007/978-981-15-1193-6_7

1 Introduction

A Palm Oil Fuel Ash (POFA) is a by-product of the palm oil mill and contains high silica content (44–66%) [1, 2]. Remarkably, it has the potential to be a Supplementary Cementing Material (SCM) in construction materials to enhance the strength, workability and durability properties of mortars and concretes [3, 4]. In the Hardened Cement Paste (HCP), silica particles from POFA undergo pozzolanic reactions with calcium hydroxide (CH) in the presence of water to generate secondary Calcium Silicate Hydrate (CSH) as a binding component [3, 4]. The nanosized-SCM (1–100 nm particles) has gained global attention as it gives a high improvement in the strength and durability properties of cement-based products [5, 6]. In that respect, the rate of cement hydrations and pozzolanic reactions has increased due to the highly specific surface area of nanosized-SCM. Furthermore, the nanosized-SCM remarkably acts as nucleation sites for the embedding of hydration products and filler to refine the microstructure of paste [5, 6].

The modification of POFA into nanosized-POFA (nPOFA) should give different perspectives on utilizing POFA as SCM in the hardened cementitious matrix. Hence, the investigation towards the cement hydration and pozzolanic reaction in the presence of nPOFA will provide a better understanding of the potential of nPOFA as SCM. The present paper examined the cement hydration extents on the Hardened Cement Paste (HCP) incorporated nPOFA via thermal and X-ray Diffraction (XRD) analysis based on the progress of cement minerals and hydration products.

2 Experimental

The experimental works of this study includes the preparation of nPOFA, the preparation of cement paste cubes, as well as the thermal and XRD analysis procedures.

2.1 Materials

As-received POFA (rPOFA) specimens were sieved through a No. 100 (150 μm) sieve to remove large contaminants. The rPOFA was subjected to the modified Los Angeles abrasion test machine to obtain a microsized-POFA (mPOFA) particle as it is compiled in the ASTM C618–15. Then, the mPOFA specimen was subjected to a ceramic ball milling for 30 h to acquire nPOFA. The mPOFA specimen was loaded into the ball milling using a 10:1 ball-to-specimen weight ratio [7].

2.2 Physicochemical Analysis

The oxide composition of nPOFA specimen and Ordinary Cement Portland (OPC) Type I was determined using an X-ray Fluorescence (XRF) spectroscopy (S4 Pioneer, Bruker). The mineralogy of nPOFA was carried out via a Siemens/Bruker Advance D5000 diffractometer with Cu Kα radiation ($\lambda = 1.54060$ Å). The analysis was operated at 40 kV and 25 mA in the range of 2θ from 2° to 90° and step sizes of 0.05° using a scanning rate of 0.05°/min at room temperature. Meanwhile, the average crystallite size (D) of nPOFA was computed using Debye–Scherrer's equation. Respectively, the specific surface area of 16.63 m^2/g and 145.35 m^2/g for the mPOFA and nPOFA specimens were evaluated from BET analysis through Thermo Fisher Scientific Sorptomatic 1990 surface analyser.

2.3 Preparation of Cement Paste Cubes

Cement paste cubes were prepared by incorporating mPOFA and nPOFA in the range of 10% w/w to 30% w/w and 10% w/w to 60% w/w, respectively. The OPC paste with water-to-cement ratio of 0.28 was used as a control paste, while the mPOFA and nPOFA cement pastes were prepared with the water-to-binder ratio of 0.35. The paste was cast in 50 mm × 50 mm × 50 mm cubic moulds for 24 h. Subsequently, the cement cubes were removed from the moulds and cured in a saturated lime solution for 28 and 90 days, respectively.

2.4 Thermal and XRD Analysis

The hydration of HCP at 28- and 90-day curing ages was monitored through thermal and XRD analysis. The HCP cubes were crushed and immersed in acetone for 3 days to arrest the hydration [8] and were kept in an oven at a temperature of 50 ± 5 °C for 24 h to allow the evaporation of the remaining acetone. A TGA was performed using a Perkin-Elmer Thermal Pyris Diamond TG/DTA analyser. The specimens were heated at a temperature of 30 °C up to 1000 °C with a constant heating rate of 10 °C min-1 under a nitrogen atmosphere. The XRD analysis was carried out using a Siemens/Bruker Advance D5000 diffractometer with Cu Kα radiation ($\lambda = 1.54060$ Å). The scan was operated at a voltage of 40 kV and a current of 25 mA in the range of 2θ from 2° to 70° and step size of 0.05° using a scanning rate of 0.05° min^{-1} at room temperature.

3 Results and Discussion

The effect of nPOFA towards cement hydration of HCP will be discussed in this section through the results from the physicochemical analysis of the materials, thermal analysis and XRD analysis of the HCP specimens.

3.1 Physicochemical Properties

Chemical composition of POFA and OPC was shown in Table 1. It is apparent that silica is the major component in POFA with 54.8%. Other compounds include aluminium oxide, calcium oxide, iron oxide, magnesium oxide and sulphur trioxide. It was also observed that POFA contains 8.5% of LOI values. Based on the oxide composition, POFA can be classified as Class-C pozzolan as it is conformable with ASTM C618-12a.

The XRD diffractogram of mPOFA and nPOFA was illustrated in Fig. 1. As can be seen, the broad peak at 20°–30° of XRD diffractograms represents the amorphous mineral in mPOFA and nPOFA. Meanwhile, the α-quartz, tridymite, cristobalite,

Table 1 Chemical composition of POFA and OPC

Component	SiO_2	CaO	Al_2O_3	MgO	Fe_2O_3	SO_3	LOI
POFA	54.8	14.0	7.24	4.14	4.47	0.71	8.5
OPC	21.45	60.98	3.62	1.22	4.89	2.30	1.37

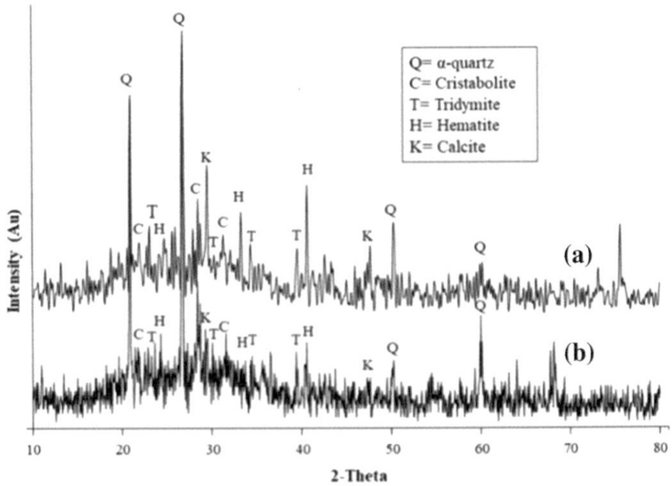

Fig. 1 Diffractogram of **a** mPOFA and **b** nPOFA

hematite and calcite peaks reflect on the crystallinity phases of POFA. The α-quartz is present as the major phase, while tridymite, cristobalite, hematite and calcite exist as minor phases of mPOFA and nPOFA. As can be seen, both mPOFA and nPOFA specimens possessed a strong and sharp peak at 26.7° which corresponds to the α-quartz peak. It is also observed that the peaks at 2θ angles of 21.7°, 50.3° and 60.4° are corresponding to α-quartz. Cristobalite peaks were identified at 2θ angles of 22.6°, 28.7° and 36.2°. The existence of tridymite peaks is identified at 2θ angles of 23.3°, 30.1°, 35.9° and 38.9°, while hematite peaks were detected at 2θ angles of 24.0°, 33.1° and 40.8°. In addition, the peaks at 29.3° and 47.3° correspond to the presence of calcite. All these peaks were similarly detected in previous studies [9].

The average crystallite size of nPOFA was found to be 34.74 nm, which was calculated based on the reflection peaks at 2θ angles of 20.9°, 26.7°, 29.4° and 50.2° by using Debye–Scherrer's equation. The findings proposed that the production of nPOFA can be achieved using a ceramic ball milling in 30 h with a high intensity of speed rotation.

3.2 Thermal Analysis

Figures 2 and 3 show the TGA curves of OPC, mPOFA and nPOFA HCP at 28- and 90-day curing ages. In this study, it is observed that three significant weight losses were detected from the TGA thermograms for all HCPs. First endotherm peak was observed in the temperature range of 30–210 °C, the second endotherm peak was detected between 410 and 480 °C, and the last endotherm peak was in the temperature range of 510–700 °C. It is suggested that the first, the second and the third endotherm

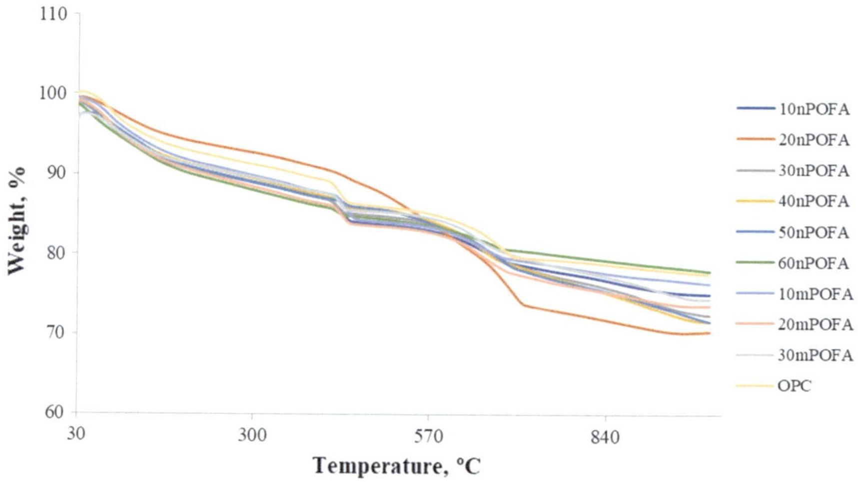

Fig. 2 TGA curves of OPC, mPOFA and nPOFA cement paste at 28-day curing age

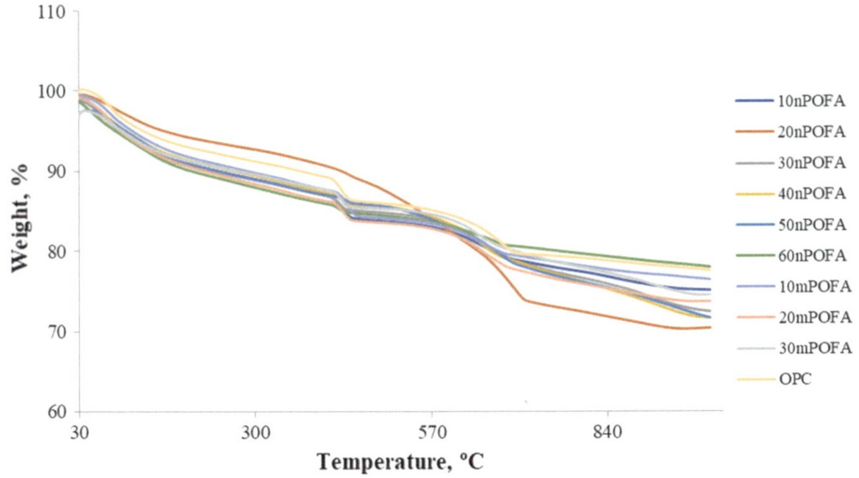

Fig. 3 TGA curves of OPC, mPOFA and nPOFA cement paste at 90-day curing age

peaks could be attributed to the decomposition of the evaporated water from capillary pores and hydrated products [10, 11], dehydroxylation of CH (shown in Eq. 1) [10] and decarbonation of carbonated phases (shown in Eq. 2) [12, 13], respectively.

$$\underset{\text{Porlandite}}{Ca(OH)_2} \xrightarrow[\text{[410 °C–480 °C]}]{} \underset{\text{Calcium oxide}}{CaO} + \underset{\text{(dehydroxylation process)}}{H_2O} \quad (1)$$

$$\underset{\text{Calcite}}{CaCO_3} \xrightarrow[\text{[510 °C–700 °C]}]{} \underset{\text{Calcium oxide}}{CaO} + \underset{\text{(decarbonation process)}}{H_2O} \quad (2)$$

Figure 4 shows the total weight of CH consumption in mPOFA and nPOFA HCPs during the time interval of 28-day curing age to 90-day curing age. As can be seen in Fig. 4, the total CH consumption of 10mPOFA, 20mPOFA and 30mPOFA HCPs within the time interval between 28-day and 90-day curing ages are 12.4%,

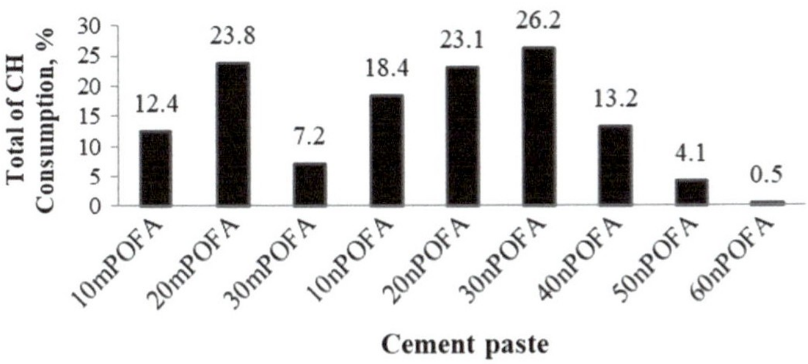

Fig. 4 Total weight of CH consumption in mPOFA and nPOFA cement pastes within the time interval of two curing ages

23.8% and 7.2%, respectively. Meanwhile, the total weights of CH consumption in the 10nPOFA, 20nPOFA, 30nPOFA, 40nPOFA, 50nPOFA and 60nPOFA HCPs are 18.4%, 23.1%, 26.2%, 13.2%, 4.1% and 0.5%, respectively. The 20mPOFA HCP shows the highest in total CH consumption among the mPOFA HCPs as the 30nPOFA HCP possesses the highest total CH consumption among nPOFA HCPs. In addition, the 30nPOFA HCP has a higher total CH consumption compared to total CH consumption in 20mPOFA HCP. Meanwhile, the lowest total CH consumption in the mPOFA and nPOFA HCPs is demonstrated in the 10mPOFA and 60nPOFA HCPs, respectively.

The result suggests that the total CH consumption indirectly assesses the degree of pozzolanic reaction in mPOFA and nPOFA HCPs. This is due to the fact that the CH is consumed by silica in the pozzolanic reaction [14, 15]. As can be seen in Fig. 4, it can be suggested that the nPOFA HCP with the cement replacement 30% w/w of nPOFA produces the high secondary CSH through pozzolanic reaction. This could be explained by the highest total CH consumption in 30nPOFA HCP. The different values in total CH consumption between 20mPOFA and 30nPOFA HCPs could be attributed to the fineness effect. The high fineness of nPOFA particle increases the filler effect, the nucleation effect, as well as the pozzolanic reaction in the hardened cementitious matrix. This agreement is similar to the previous findings [16, 17]. Meanwhile, the low CH consumption in the HCP with the high composition of mPOFA or nPOFA could be attributed to the high dilution effect as it was mentioned by the previous study [15].

3.3 XRD Analysis

Figure 5 shows the diffractogram of OPC, mPOFA and nPOFA HCPs at 28- and 90-day curing ages, respectively. It is observed that the OPC HCP has the highest peak intensity of crystalline CH compared to mPOFA and nPOFA HCPs at 28- and 90-day curing ages. As can be seen in Fig. 5b, the CH peaks at $2\theta = 18.1^\circ$ and 34.0° in 30nPOFA and 40nPOFA HCPs at 90-day curing age show the lower intensities compared to the other HCPs. In addition, it can be seen in Fig. 5 that the α-quartz peaks at $2\theta = 26.5^\circ$ were not observed in OPC HCP as it was observed in all mPOFA and nPOFA HCPs.

The result suggests that the pozzolanic reaction in the mPOFA and nPOFA HCPs results in the low intensity of CH peaks, which were observed in diffractogram. The results were confirmed by the thermal analysis, which suggests that the high CH consumption in mPOFA and nPOFA HCPs (refer to Fig. 4) could be attributed to the pozzolanic reaction. This is in agreement with the results reported by other researchers [15]. The low intensity of CH peaks at $2\theta = 18.1^\circ$ and 34.0° in 30nPOFA and 40nPOFA HCPs at 90-day curing age is likely to be due to the high rate of pozzolanic reaction in 30nPOFA and 40nPOFA HCPs. It is confirmed by the high

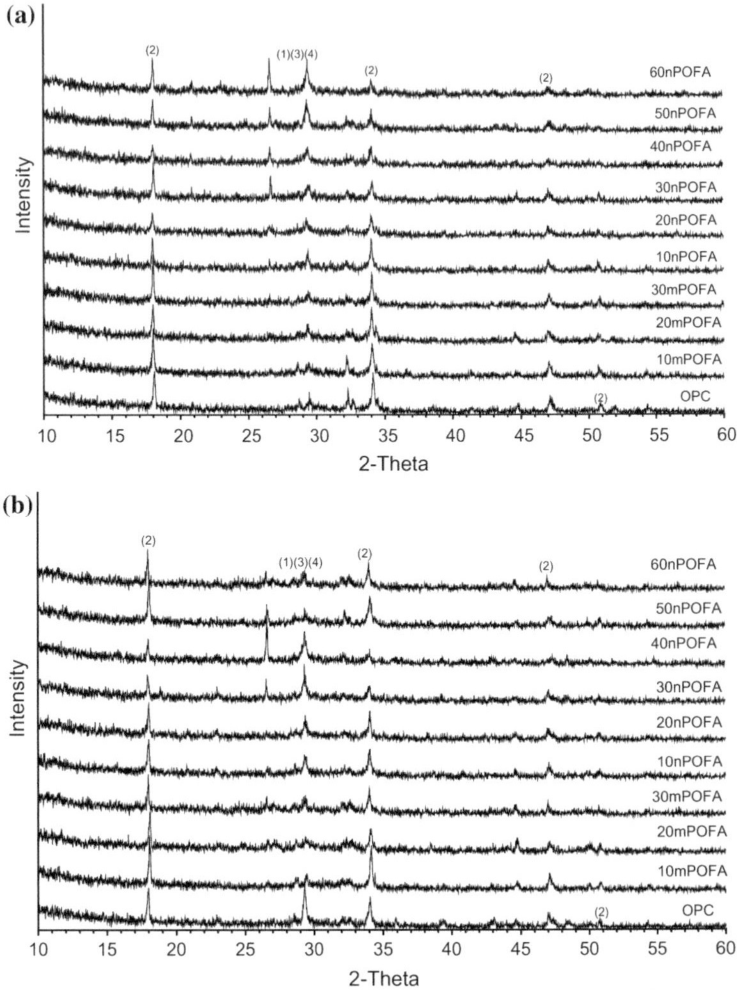

Fig. 5 Diffractogram of OPC, mPOFA and nPOFA HCPs at **a** 28-day curing age, **b** 90-day curing age. Notation—(1) Alite, (2) Portlandite, (3) Belite, (4) Calcite

CH consumption (refer to Fig. 4) in HCP. It can be seen in Fig. 5b that the CH peaks at 2θ = 34.0° in nPOFA HCP, which is significantly lower compared to the peaks in mPOFA and OPC HCPs. This could be attributed to the high fineness of nPOFA, which increases the rate of pozzolanic reaction in HCP.

4 Conclusions

The findings of the present study can be summarized as follows:

(1) The TGA analysis investigates the hydration progress of HCP incorporated mPOFA and nPOFA based on the weight loss of CH. The findings show that the 30nPOFA has the highest CH loss in the HCP compared to the other HCPs, as it reflects on the highest CH consumption in the 30nPOFA within the interval time of two curing ages. The result indicates that the cement replacement with 30% w/w of nPOFA could give the optimum pozzolanic reaction to reduce CH amount in HCP.

(2) The XRD analysis was carried out to qualitatively investigate the hydration progress based on the crystalline phases in HCP. The significant reduction in the intensity of CH peaks in the nPOFA HCP compared to the mPOFA could be attributed to the high rate of pozzolanic reaction in HCP.

Acknowledgements The authors thank the staff of Department of Chemistry, Faculty of Science and the Department of Structure and Materials, Faculty of Civil Engineering, Universiti Teknologi Malaysia. The authors also acknowledge Universiti Malaysia Sabah and Ministry of High Education Malaysia for the financial support.

References

1. Abdullah K, Hussin MH, Zakaria M et al (2006) POFA: a potential partial cement replacement material in aerated concrete. In: 6th Asia Pacific structural engineering and construction (APSEC 2006), 5–6 September. https://doi.org/10.4028/www.scientific.net/AMM.567.446
2. Tangchirapat W, Khamklai S, Jaturapitakkul C (2012) Use of ground palm oil fuel ash to improve strength, sulfate resistance, and water permeability of concrete containing high amount of recycled concrete aggregates. Mater Des 41:150–157. https://doi.org/10.1016/j.matdes.2012.04.054
3. Al-Mulali MZ et al (2015) The incorporation of oil palm ash in concrete as a means of recycling: a review. Cem Concr Compos 55:129–138. https://doi.org/10.1016/j.cemconcomp.2014.09.007
4. Safiuddin M, Salam MA, Jumaat MZ (2013) Utilization of palm oil fuel ash in concrete: a review. J Civ Eng Manag 17(2):234–247. https://doi.org/10.3846/13923730.2011.574450
5. Singh LP et al (2013) Beneficial role of nanosilica in cement-based materials: a review. Constr Build Mater 47:1069–1077. https://doi.org/10.1016/j.conbuildmat.2013.05.052
6. Al-Jabri K, Shoukry H (2014) Use of nano-structured waste materials for improving mechanical, physical and structural properties of cement mortar. Constr Build Mater 73:636–644. https://doi.org/10.1016/j.conbuildmat.2014.10.004
7. Abdul Khalil HPS et al (2011) Preparation and characterization of nanostructured materials from oil palm ash: a bioagricultural waste from oil palm mill. BioResources 6(4):4537–4546
8. Zhang J, Scherer GW (2011) Comparison of methods for arresting hydration of cement. Cem and Concr Res 41(10):1024–1036. https://doi.org/10.1016/j.cemconres.2011.06.003
9. Megat Johari MA et al (2012) Engineering and transport properties of high-strength green concrete containing high volume of ultrafine palm oil fuel ash. Constr Build Mater 30:281–288. https://doi.org/10.1016/j.conbuildmat.2011.12.007
10. El-Diadamony H et al (2016) Hydration and characteristics of metakaolin pozzolanic cement pastes. HBRC J 14(2):150–158. https://doi.org/10.1016/j.hbrcj.2015.05.005

11. Deboucha et al (2017) Hydration development of mineral additives blended cement using thermogravimetric analysis (TGA): methodology of calculating the degree of hydration. Constr Build Mater 146:687–701. https://doi.org/10.1016/j.conbuildmat.2017.04.132
12. Morsy MS, Alsayed SH, Aqel M (2010) Effect of nano-clay on mechanical properties and microstructure of ordinary portland cement mortar. Int J Civ Environ Eng 10(1):23–27
13. Saikia N, Kato S, Kojima T (2006) Thermogravimetric investigation on the chloride binding behaviour of MK-lime paste. Thermochim Acta 444(1):16–25. https://doi.org/10.1016/j.tca.2006.02.012
14. Rajak MAA, Majid ZA, Ismail M (2019) Pozzolanic activity of nanosized palm oil fuel ash: a comparative assessment with various fineness of palm oil fuel ash. IOP Conf Ser Earth Environ Sci 220:012061. https://doi.org/10.1088/1755-1315/220/1/012061
15. Siddique R, Khan MI (2011) Supplementary cementing material. Springer-Verlag, Berlin Heidelberg
16. Rajak MAA, Majid ZA, Ismail M (2015) Morphological characteristics of hardened cement pastes incorporating nano-palm oil fuel ash. Procedia Manuf 2:512–518. https://doi.org/10.1016/j.promfg.2015.07.088
17. Pacheco-Torgal F, Jalali S (2011) Nanotechnology: advantages and drawbacks in the field of construction and building materials. Constr Build Mater 25(2):582–590. https://doi.org/10.1016/j.conbuildmat.2010.07.009

Changes in the Use of Building Materials in Traditional Houses in Indonesia

Andi Prasetiyo Wibowo

Abstract The residence or what we call home is an inseparable part of human life. Traditional buildings in Indonesia use natural materials that are widely available in Indonesia, such as wood, bamboo, stone, clay, straw, weeds, and so on. The use of natural materials for traditional houses will certainly have a negative effect if the exploitation of natural materials is carried out without being balanced with policies for the renewal process. Modern society tends to look for substitutes for building materials that are simpler, or at least easy to obtain, and that do not require special expertise in the installation/manufacturing process. Comparisons between traditional and modern versions of traditional houses that have been touched by modern materials are expected to give a picture of a shift in the types of building materials and their application to traditional houses in Indonesia.

Keywords Houses · Traditional · Indonesia · Building materials

1 Introduction

The residence, which later on we will call it home, is an inseparable part of human life. Home is a basic human need as a place to rest and shelter from the weather and the threat of the surrounding environment that can endanger him. In the process of this development, humans initially only relied entirely on nature, such as searching for caves or taking shelter under large trees, up to a time when humans were able to process the natural resources to use them to build houses, starting from a simple scale to those who applied technology in the manufacturing process.

The process of making houses in various regions of the world has also an indirect impact on the development of culture with the birth of the types of traditional houses from various regions that can give a characteristic of certain regions. The use of natural materials available in the area is a rational choice when people want to build their houses. In addition to being easy to get, at that time, relations with other regions

A. P. Wibowo (✉)
Department of Architecture, Faculty of Engineering, Universitas Atma Jaya Yogyakarta, Depok, Indonesia
e-mail: andi.prasetiyo@uajy.ac.id

M. Awang and M. R. Meor M Fared (eds.), *ICACE 2019*,
Lecture Notes in Civil Engineering 59, https://doi.org/10.1007/978-981-15-1193-6_8

that had other types of material were certainly still very limited. The character of the traditional houses also characterizes the natural and geographical conditions of the area. Customs are also very influential in determining the design of the house in order to maintain its locality, as happened in kaampong Naga Tasikmalaya, West Java [1].

The use of natural materials for traditional houses will certainly have a negative effect if the exploitation of natural materials is carried out without being balanced with policies for the renewal process. The design of the building turned out to also have a big contribution to trigger environmental damage and result in a decrease in the quality of human life [2]. Although the natural ingredients are categorized as renewable materials, the number of natural resources that are increasingly getting limited and the shrinking of land for the renewal process threaten the continuity and availability of these building materials. According to the economic principle, if the amount is limited and needs increasing, it will make the price of the building materials expensive. Generally, the price of building materials causes housing prices to become more expensive, and at some point, people will choose to create housing that may use as little as possible natural materials.

The same thing happened in Indonesia. As an archipelago consisting of many tribes, the Indonesian nation has many extraordinary vernacular buildings [3]. Traditional buildings in Indonesia use natural materials that are widely available in Indonesia, such as wood, bamboo, stone, clay, straw, weeds, and so on. The use of these natural materials in addition to characterizing the locality of the area is also able to provide additional information about the progress of the level of thinking of the local community through the processing of natural materials to be used as housing material. The processing and utilization of natural materials can be different between one region and another, and therefore it can create diversity in its application. These traditional houses will later become a source of cultural treasures from the Indonesian state, which consists of various ethnic groups. These traditional houses are also a part of the development of national culture.

In almost all regions in Indonesia, the traditional houses also have a role to show the social status of the residents. The types of traditional houses have levels ranging from simple to relatively large, and they are filled with beautiful symbols or architectural details. Traditional houses for nobles—someone who has a respectable position in the region—tend to have complicated constructions and large sizes, so that it requires a lot of building materials. These things have an impact on the length of the process of making the traditional house. Not to mention the need for special artisans because there are several parts of the house that require certain skills for the technique of making/installing some elements of the traditional house.

Houses, as assets, will eventually become an integral part of the regeneration process of mankind. A house will be inherited from one generation to the next. Over time, of course, the house will get old and needs care. The limited resources (both material and labor) ultimately made the majority of the community find solutions to these problems by finding a replacement for building materials that were simpler, or at least easier to obtain, and did not require special expertise in the installation/manufacturing process. As has been studied in Iran, physical and morphological changes in a house can also be triggered by social, economic, and technological transformation [4].

2 Research Methodology

This study will use a comparative method between several types of traditional buildings in Indonesia. Comparisons are made among traditional houses, original versions, and modern versions that have been touched by modern materials. The analysis process uses primary data obtained in the field by conducting observations and taking documentations, while secondary data is obtained from literature studies through some traditional home documentations obtained from various library sources.

3 Analysis and Discussion

The analysis process is done by comparing traditional houses in terms of building components. The observed part will be identified regarding the building material structure. Some traditional houses that will be compared are taken from several examples, which are expected to be able to represent regions in Indonesia, starting from the islands of Sumatra, Kalimantan, Java, Bali, Sulawesi, Nusa Tenggara, and Papua.

- *Rumah Gadang* (Minangkabau—West Sumatera)

Rumah Gadang (Gadang House) is the name for the Minangkabau traditional house, which is a traditional house and is often found in the province of West Sumatra, Indonesia. This house is also called by other names by the local community such as *Rumah Bagonjong* or there is also someone who calls it the Baanjuang house (Figs. 1 and 2).

- Betang House (Kalimantan)

Fig. 1 Original version of the Minangkabau house

Fig. 2 The recent and most common Rumah Minangkabau

Fig. 3 Original Betang house (left) and current (right)

Rumah Betang is a typical Kalimantan traditional house which is found in various parts of Kalimantan and is inhabited by Dayak people, especially in the upper reaches of the river which is usually the center of Dayak tribal settlements. The most significant change that has taken place is the replacement of the type of roofing material that originally used shingles, and now it has been replaced with a zinc roof (Fig. 3).

- Rumah Joglo (Jawa)

Joglo house uses wood as its structure and uses woven bamboo as its walls, while for roofs, it uses sago palm before finally being replaced with tile or precarious clay (Fig. 4).

- Bali

The traditional Balinese house originally used palm fiber roofing and wood support, and for walls, "popolan" or clay soil was used. In its development, due to the impact of the advancement of tourism, many Balinese people then tried to transform the building into tourist facilities for tourists [5]. One part that is undergoing rejuvenation is that the roof of the building is replaced with a metal/zinc roof (though not entirely), and a floor covering that uses tiles or stones (Fig. 5).

- Minahasa (North Sulawesi)

Traditional Minahasa house is stage house with 16–18 supporting poles and twin stairs on the front of the building (Fig. 6).

Fig. 4 Original Joglo house (left) and current (right)

Fig. 5 Original Balinese traditional house (left) and current (right)

Fig. 6 Original Minahasa traditional house (left) and current (right)

Fig. 7 Use of concrete stairs instead of wooden stairs

The roof of the Minahasa building in the past used material in the form of sago palm which was called sago tree. The current condition of Minahasa's roof has begun to change by using zinc material due to the difficulty in getting rumbia and its declining quality. Apart from material problems, there are also other factors where roof construction also changes (Fig. 7).

- Toraja (South Sulawesi)

Tongkonan is a traditional house of the Toraja people. The roof is curved to resemble a boat, consisting of a composition of bamboo trees and at this time some Tongkonan have been replaced using a zinc roof. One characteristic of the Tongkonan house, which is on the front, is that there is a row of buffalo horns that indicate the social status of the owner (Fig. 8).

- West Nusa Tenggara (Sasak tribe, Sade village, Lombok)

Bale is a traditional house from the Sasak tribe in Sade hamlet in Rembitan village, Pujut, Central Lombok. The uniqueness of Sade hamlet is its determination to preserve this traditional house. The use of roofing material from straw is maintained

Fig. 8 Original Toraja traditional house (left) and current (right)

Fig. 9 Traditional Sasak tribal house, Bale Lumbung (left) and Bale Tani (right)

Fig. 10 Modernization of Bale Lumbung house

in traditional house buildings in remote/traditional villages. But for new buildings, it tends to use new materials such as metal roofs and construction of wooden walls with glass (Figs. 9 and 10).

- Papua

Papua traditional house is better known as HONAI. Typical roof forms and the use of natural materials such as wood for walls, and the use of straw and palm fiber as roof cover materials, make the traditional house of Papua region easily recognizable. However, due to the demands of health standards, *the honai* currently uses concrete floors (Fig. 11).

Fig. 11 Original Papuan traditional house (left) and current (right)

4 Conclusions

The most visible building identity is the shape of the roof. The shape of the roof of a traditional house has not changed because it is a symbol/characteristic of the traditional house itself. However, many of the constituent materials change from sago palm, palm fiber, bamboo, thatch becomes precarious soil, zinc, and tile metal.

For walls, not much has changed and they still use wood or bamboo, but they are equipped with glass windows.

The floor section of the original version of the traditional house uses wooden boards for the stage house. However, in its development, many tread houses were changed using concrete pavement and using ceramic floors.

For the foundation or the bottom part of the building, the initial use of wood as its supporting structure is then replaced by concrete.

References

1. Ningrum E (2017) Dinamika Masyarakat Tradisional Kampung Naga di Kabupaten Tasikmalaya. MIMBAR J Sos dan Pembang 28(1):47
2. Wibowo, AP (2014) Kriteria Rumah Ramah Lingkungan (eco-friendly house). J Muara Sains Teknol Kedokt dan Ilmu Kesehat 1(1):1–10
3. Dawson B, Gillow J (1994) The traditional architecture of Indonesia. Thames and Hudson, London
4. Mirmoghtadaee M (2009) Process of housing transformation in Iran. J Constr Dev Ctries 14(1):69–80
5. Bobrovskaya L et al (2013) Does exposure to chronic stress influence blood pressure in rats? Auton Neurosci Basic Clin 177(2):217–223

Fatigue and Long-Term Deformation Performance of Conventional and Rubber-Modified Bitumen

Mahirah Mohd Idris, Madzlan Napiah, Mohd Yazip Matori and Nazirah Ahmad

Abstract The intention of road pavement is to provide a convenient top surface for vehicles so that they can easily move without any difficulties, and safely from one place to another. The flexible pavement using conventional bitumen usually shows defects such as rutting, cracking, and fatigue failure, thus causing damage to roads earlier than the expected time. Therefore, an approach to bitumen modification using rubber has been done, so that it is more resistant to any deterioration of the pavement performance. An integration of natural rubber and bituminous materials is a tremendous innovation in world road pavement. Generally, natural rubber has high tensile strength and it is resistant to fatigue due to its elastic substance which is known as its feature. This modification is able to give an encouraging effect to the resilience of the road compared to the conventional bitumen, which has a limitation toward traffic loading. A consolidation of 6% of natural rubber to the 60/70 grade penetration of bitumen has shown an improvement to its main property. As we know, conventional bitumen susceptibility is directly proportional to the increasing of the traffic loading. With this alteration using a chemical composition in conventional bitumen is expected to prolong the lifespan of road pavement. The behavior of the modifier (natural rubber) has been observed by inclusive laboratory testing and appraisal.

Keywords Natural rubber · Modified bitumen · Pavement performance

M. M. Idris (✉) · M. Napiah
Department of Civil & Environmental Engineering, Universiti Teknologi PETRONAS, 31750 Tronoh, Perak, Malaysia
e-mail: mahirahmohdidris@gmail.com

M. Napiah
e-mail: madzlan_napiah@utp.edu.my

M. Y. Matori
IKRAM Pavement Laboratory, Block 8, Unipark Suria, Jalan IKRAM-Uniten, 43000 Kajang, Selangor, Malaysia
e-mail: yazip@ikram.com.my

N. Ahmad
Engineering & Technology Division, Malaysian Rubber Board, 47000 Sungai Buloh, Selangor, Malaysia
e-mail: nazirah@lgm.gov.my

M. Awang and M. R. Meor M Fared (eds.), *ICACE 2019*,
Lecture Notes in Civil Engineering 59, https://doi.org/10.1007/978-981-15-1193-6_9

1 Introduction

There is no hesitation that our contemporary road structures are necessitated to have better long-term performance and resist higher traffic densities and axle loadings compared to the past. At the same time, the emphasis of reducing material costs is also given attention. In order to ensure long-term durability, and at the same time, minimising maintenance measures and conserving resources, proper selection of paving materials together with the optimal mix and pavement design are very crucial. Basically, bituminous binders should be designed to provide optimum resistance to different pavement failures and to achieve great durability within a diversity of traffic conditions. This, however, is not what has usually happened with conventional bitumen.

Bitumen modification is one of the steps that can be considered to improve the performance of asphaltic pavement, especially in sustainability of its lifespan. Al-ani, 2009 [1] has identified that the bitumen modification with polymer is an initiative to enhance road pavements so that it is excellent in terms of performance durability. In this case, natural rubber is selected to be as a modifier to the bitumen in order to increase the viscosity and elasticity [2] of asphalt with higher softening point, penetration, and ageing resistance. These authentications can be found at the rheological testing, which has been done at the earlier stages of this investigation. Furthermore, with this great effect in rheological properties, a distress reduction in the asphalt pavement can be expected to reduce distress in asphalt pavement. In addition, natural rubber also has high stability, excellent tear strength, and fatigue resistance, which can extend the durability in asphalt pavement [3]. Hence, the presence of natural rubber as a modifier element in bitumen modification for road construction is technically to lengthen the lifetime of roads and eventually save the maintenance cost [4].

2 Research Problem

The utilization of conventional bitumen in pavement as a binder in aggregates had been generally known. Unfortunately, the development of the world has further challenged the roads with the increments in traffic densities and the expansion in axle load, which bring the encumbrance to the pavement. In long run, the pavement cannot accommodate the higher traffic volumes, which finally lead to the deterioration of the road.

3 Research Background

Bitumen is known as the oldest engineering material [5]. Bitumen had many usages such as adhesive and sealant. The most popular usage of bitumen is commonly in pavement binder. Unfortunately, bitumen is always a challenge to get good quality. The properties of the bitumen are associated closely to the process of refinery as well as to the crude oil sources. Hence, the chemical composition of bitumen is allegedly various and complex. Having the quality of crude oil and well quality control during refinery are the correct ways to preserve the continuity of good resources. However, it is hard to maintain this method as the good quality of crude oil has its limitations and lacks effectiveness during the refining process. Therefore, bitumen modification has been introduced to overcome this issue and to sustain bitumen's consumption in order to preserve the economy [4]. Many researchers had focused on this perspective to improve the quality of bitumen by doing modifications through its chemical and mechanical properties, using polymers such as plastomers and thermoplastic elastomers. However, there are still some challenges and drawbacks toward polymer modification [6]. In this study, natural rubber has been used to be mixed with bitumen because of its good interaction in terms of physical and chemical properties to the nonrubber substances [7]. For its physical properties, natural rubber has elastic behavior, which can be seen when it is under loading, and it can return to its original condition when the loading is removed [8]. In addition, natural rubber also unveils that it can lower its elasticity value while it is stretched [9]. The chemical formula of rubber shows that it is composed of carbon and hydrogen atoms (C5H8) with higher molecular weight. The presence of the double bond in this formula for each unit makes the rubber increase its elasticity [10].

Instead of improving the performance of the pavement, the utilization of natural rubber as a modifier element can also increase the domestic consumption of rubber, especially to the biggest natural rubber producers such as Thailand, Malaysia, and Indonesia [11].

A. Research Hypothesis

The concept of using rubber in bitumen has been already known where rubber crumb was widely used in rubber-bitumen modification. However, nowadays the approach of using natural rubber is still new. In this investigation, natural rubber with 80% dryness has been used. This approach is able to overcome the road problems, such as ageing, rutting, and fatigue failure due to various load vehicles.

B. Research Objectives

The objective of this research is to evaluate the performance of the pavement for conventional and modified bitumen using natural rubber, based on comprehensive laboratory testings such as wheel tracking test, indirect tensile modulus test, and asphalt permanent deformation test.

4 Research Methodology

The Optimum Binder Content (OBC) for both types of bitumen was retrieved from the previous experiments which are Stability and Flow Tests. The OBC 60/70 penetration grades of bitumen and modified bitumen using natural rubber are as follows.

A. Wheel Tracking Test

Two rectangular samples of slabs with different types of bitumen were prepared for evaluation of rutting in laboratory. The OBC for both samples were used as in Tables 1 and 2, respectively. The compaction for both samples was done at the temperature of 150 °C for modified bitumen and 135 °C for conventional bitumen. The samples were placed to rest for 24 h prior to testing. After 24 h of resting, the samples were then put individually in the wheel tracking machine. The machine tracks a loaded steel wheel back and forth directly on the sample for 40 min (Figs. 1 and 2).

B. Indirect Tensile Modulus Test

Three specimens for each type of bitumen were prepared. A cyclic stress of fixed amplitude is applied to the test specimens. During testing, the specimens are subjected to a dynamic cyclic stress and a seating load. Loads are selected to keep horizontal deformations. The deformation responses of the specimens were measured near to the center of the specimen. Total resilient modulus is calculated using the recoverable horizontal deformation. The test for each specimen was performed at a temperature of 25 °C. The average value of stiffness modulus (MPa) is taken (Fig. 3).

Table 1 The OBC for 60/70 penetration grade and modified bitumen

Marshall properties	% bitumen at optimum curve	
	60/70	Modified bitumen
Density (kg/m3)	5.50	5.50
Stability (N)	4.50	4.50
Air voids	4.80	5.45
% voids filled with bitumen	4.90	5.40
Flow value (mm)	4.80	5.00
Optimum bitumen content	4.90	5.20

Table 2 Results for both types of bitumen

Type of bitumen	Wheel tracking (mm)	Stiffness modulus (Mpa)	Min strain rate (s^{-1})
60/70	4.30	139.50	1.15
Natural rubber modified bitumen	2.80	365.80	0.18

Fig. 1 The slabs after compaction

Fig. 2 The slabs in the machine

C. Asphalt Permanent Deformation Test

As in B, three samples were also prepared for both types of bitumen. Each sample was tested using the same machine as indirect tensile modulus test at a temperature of 25 °C. In this experiment, the samples were damaged because of the deformation of the samples, as the number of cycles increased. At this point, the minimum strain value can be achieved. The average value of minimum strain rate (s^{-1}) is taken.

5 Results and Discussions

The following data show the performance results for conventional and modified bitumen, respectively.

Fig. 3 The machine for indirect tensile modulus test

For wheel tracking results, it was observed that the conventional bitumen was deepened by 4.3 mm, while the natural rubber modified bitumen has only 2.8 mm depth after 40 min of testing. The distress of the pavements obviously occurs for the conventional bitumen with the emergence of rutting on top surface of the slab sample with more than 4.0 mm. According to Jabatan Kerja Raya (JKR) standard, the acceptable maximum depth for rutting is 4.0 mm. The depth of rutting for modified bitumen is still under the limit. By looking at this result, we can say that the modification of bitumen can increase the strength and enhance its durability.

The advantage of modified bitumen can also be seen through the stiffness modulus and strain rate. We can see that the modified bitumen has more than double stiffness modulus compared to the conventional bitumen. Same goes to the minimum strain rate where the value for modified bitumen is less than one-sixth compared to the conventional bitumen. It shows that, the modified bitumen is harder and more resilient to the higher loading, and thus provides better performance and services to the asphalt pavement.

6 Conclusion

A modification toward asphalt binder with natural rubber is an alternative way to enhance the quality of the pavement. The performance results from the experiment indicated that the modified binder was more sustained as compared to the conventional asphalt mixture. Good effects can be seen in the rubber modified bitumen through rutting, stiffness, and strain testings where the values given are really promising as compared to the conventional bitumen. These results are depicted to the actual road performance under a genuine condition.

Acknowledgements The author would like to acknowledge Universiti Teknologi PETRONAS, IKRAM Pavement Laboratory, and Malaysian Rubber Board for the continuous support in terms of facilities and assistantship. Thankful appreciation is also extended to Malaysian Rubber Board for its contribution in financial support for this study.

References

1. Al-ani TMA (2009) Modification of asphalt mixture performance by rubber-silicone additive. Anbar J Eng Sci 2(1)
2. Al-Mansob RA, Ismail A, Alduri AN, Azhari CH, Karim MR, Yusoff NIM (2014) Physical and rheological properties of epoxidized natural rubber modified bitumens. Constr Build Mater 63:242–248
3. Tayfur S, Ozen H, Aksoy A (2007) Investigation of rutting performance of asphalt mixtures containing polymer modifiers. Constr Build Mater 21(2):328–337
4. Polacco G, Berlincioni S, Biondi D, Stastna J, Zanzotto L (2005) Asphalt modification with different polyethylenebased polymers. Eur Polymer J 41(12):2831–2844
5. Morgan P, Mulder A (1995) The shell bitumen industrial handbook. Surrey, Shell Bitumen
6. Zhu J, Birgisson B, Kringos N (2014) Polymer modification of bitumen: advances and challenges. Eur Polymer J 54:18–38
7. Thomas S, Chan C, Othen L, Joy J, Maria H (2013) Composites and nanocomposites. In: Natural rubber materials, vol 2. Royal Society of Chemistry, Cambridge
8. Mark JE (ed) (2007) Physical properties of polymers handbook. AIP Press, New York
9. Kohjiya S, Ikeda Y (eds) (2014) Chemistry, manufacture and applications of natural rubber. Elsevier
10. Krevelen DWV, Nijenhuis KT (2009) Properties of polymers: their correlation with chemical structure; their numerical estimation and prediction from additive group contribution. Elsevier
11. Nrachai T, Chayatan P, Direk L (2005) The modification of asphalt with natural rubber latex. Proc East Asia Soc Transp Stud 5:679–694

Structural Behaviour of Strengthened High-Performance Concrete-Filled Double Skin Steel Tube Beams

Ehsan Nikbakht, Nur Illia Shuhadah, Ramanathan A/L Ganapathy, Shukanthi A/P Subramaniam and Nabilah Abu Bakar

Abstract Recently, the use of concrete-filled steel tube (CFST) composite beams has increased in the modern structural projects. In this study, the bending behaviour of high performance concrete-filled double skin tube (HPCFDST) strengthened with different CFRP layers is investigated. Mechanical properties of the incorporated high-performance self-compacting concrete are investigated. The structural performance of HPCFDST beam without CFRP layer as the control beam is compared with the HPCFDST beams strengthened with two and three layers of CFRP. The results show that the beam with CFRP sheets exhibited 19% greater load carrying capacity compared to the control beam. The beam with three layers of CFRP showed approximately similar strength, however, it exhibited greater ductility and deformation capacity compared to the beam with two layers of CFRP.

Keywords Composite structures · Double-skin steel tube · Beams · High performance concrete · CFRP

1 Introduction

The use of concrete-filled steel tube (CFST) composite beams has increased exponentially over the past few years. In the concrete-filled steel tube (CFST) members known as composite sections, which are a combination of concrete and steel, prevent buckling of steel before the yielding and enable them to achieve full advantages of both steel and concrete. To date, many researchers have conducted experimental and analytical investigations on CFSTs in various forms, such as rectangular and circular cross sections [1–4]. In recent years, concrete-filled double skin tube beams (CFDST), which provide similar advantages as CFST but with lighter weight have been introduced to industry [5]. CFDSTs are hollow section composite members with

E. Nikbakht (✉) · N. I. Shuhadah · Ramanathan A/L Ganapathy · Shukanthi A/P Subramaniam
Civil and Environmental Engineering Department, Universiti Teknologi PETRONAS, Seri Iskandar, Malaysia
e-mail: eshan.nikbakht@utp.edu.my

N. A. Bakar
Civil Engineering Department, Universiti Putra Malaysia (UPM), Seri Kembangan, Malaysia

M. Awang and M. R. Meor M Fared (eds.), *ICACE 2019*,
Lecture Notes in Civil Engineering 59, https://doi.org/10.1007/978-981-15-1193-6_10

concrete filled between inner and outer steel tubes. They are usually being applied as major structural members such as columns and beams because of their numerous advantages such as high strength, damping and ductility [6–9].

The carbon fibre-reinforced polymer (CFRP) has been used to wrap the CFST to enhance the performance of CFST in terms of the moment capacity, energy absorption and flexural stiffness [10–12]. In this study, CFRP is used to strengthen the CFDST beams filled with high performance concrete (HPC). CFRP may control the local outward buckling of CFDST, while the inward buckling is controlled by HPC. To date, there are still limited studies conducted on HPC-filled double skin steel tube beams. Hence, the aim of this research is to investigate the performance of HPCFDST with different number of CFRP layers. In this study, high performance self-compacting concrete without coarse aggregates incorporated as a core between inner and outer steel tubes are examined. A control beam without CFRP and two beams with two and three layers of CFRP are investigated. Finally, the bending behaviour and failure modes of the beams in terms of stiffness, strength and ductility are investigated.

2 Experimental Work

All experimented beams in this study have the length of 1.5 m. The cross-sectional size of the outer steel tubes is 75 mm × 100 mm and the inner tubes is 25 mm × 50 mm. The thickness of the steel tubes is 2.3 mm. The mixture design of the concrete used is shown in Table 1. The concrete has a compressive strength of 85 MPa. Further information regarding the mechanical properties of HPC used in this study can be seen in the earlier works [13]. In this study, the CFRP layers will be varied from two to three layers. The wrapping scheme will be partial and unidirectional. The adhesive material was used to bond the CFRP sheets with steel tubes with the ratio of 4:1 by weight of resin and hardener.

The three HPCFDST beams are subjected to four-point bending test. The strengthening schemes are conducted by partially applying CFRP sheets to the beam as shown

Table 1 Mixture design of HPC used in this study

Ingredient (kg/m^3)	Mix design
Portland cement	600
Fly ash	390
Silica fume	78.75
Water	194.25
River sand	480
Superplasticizers	45.15
Steel fibre	21.37
Quartz sand	154.35
Water/binder ratio	0.18

Table 2 Beam specimens and the labelling

Beam labelling	Number of CFRP layers	Diagram
B-C	–	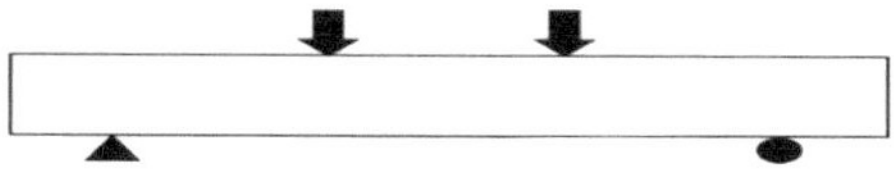
B-2L	2	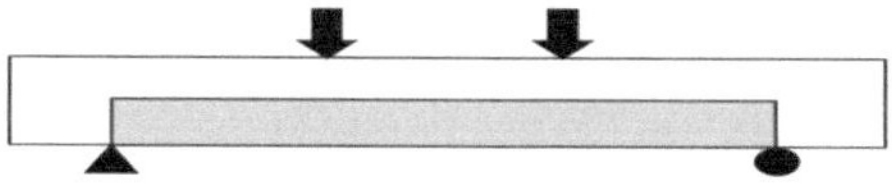
B-3L	3	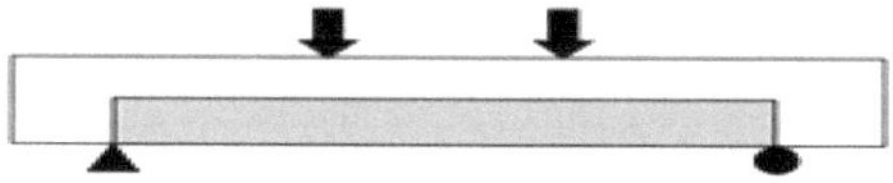

in Table 2. In addition, the labelling of the beams is stated in the table: 'B' is beam, and '2L' and '3L' represent the number of CFRP layers used. The beam surface was cleaned with sand paper, grinding machine and acetone prior to bonding of CFRP layers in order to ensure the beam is free from rust and painting. Three linear variable displacement transducers (LVDTs) were used to measure the displacement of the beams.

3 Results and Discussions

The load-mid-span displacement graphs of the experimented beams are shown in Fig. 1. As can be seen from the figure, among all the beams, the beam B-3L shows the highest load capacity of 64 kN, whereas the control beam without CFRP layer (B-C) obtained the lowest strength of 51 kN, which demonstrates that CFRP layers enhance the load capacity of the HPCFDST beams. However, as the results illustrate, the beams B-2L and B-3L with two layers and three layers of CFRP, respectively, showed approximately similar load capacities, however, the beam B-3L exhibited greater ductility compared to B-2L. The reason for the less ductile behaviour of B-2L compared to the control beam and the B-3L is due to the crushing of core concrete before the yielding of steel tubes. As a result, the beam B-2L shows lower deformation capacity where it fails at the mid-span with displacement of 40 mm as shown in Fig. 2. The graphs in Fig. 2 are obtained from LVDT's readings where the LVDTs are located at the bottom of the beams. The position of LVDTs is at the middle of the beam, and the distance of 325 mm from both supports. As can be seen from figure, the control beam has the highest deflection, while B-2L has the least deflection. Moreover, the control beam and B-3L showed approximately similar deflection capacities, where they failed at the displacements of 72 mm. In addition, as depicted in Fig. 5, the CFRP sheets on the surface of the beam B-2L

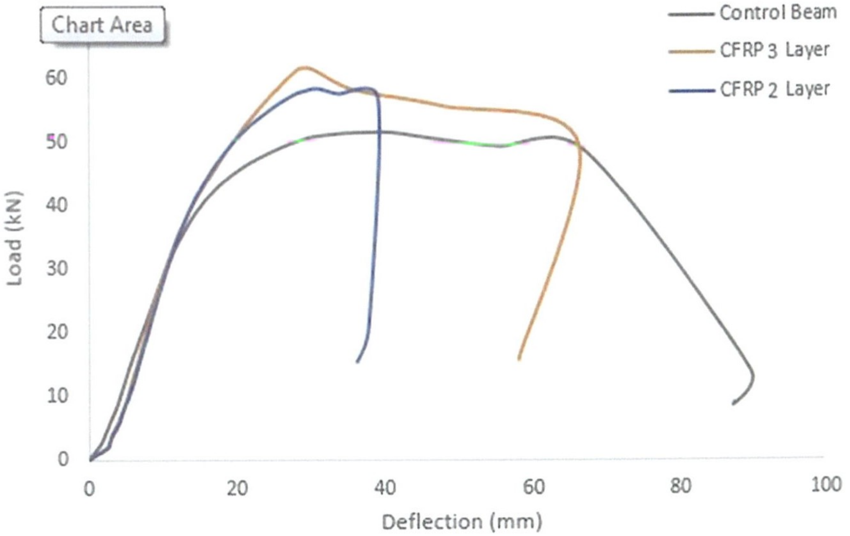

Fig. 1 Load–displacement response of the beams

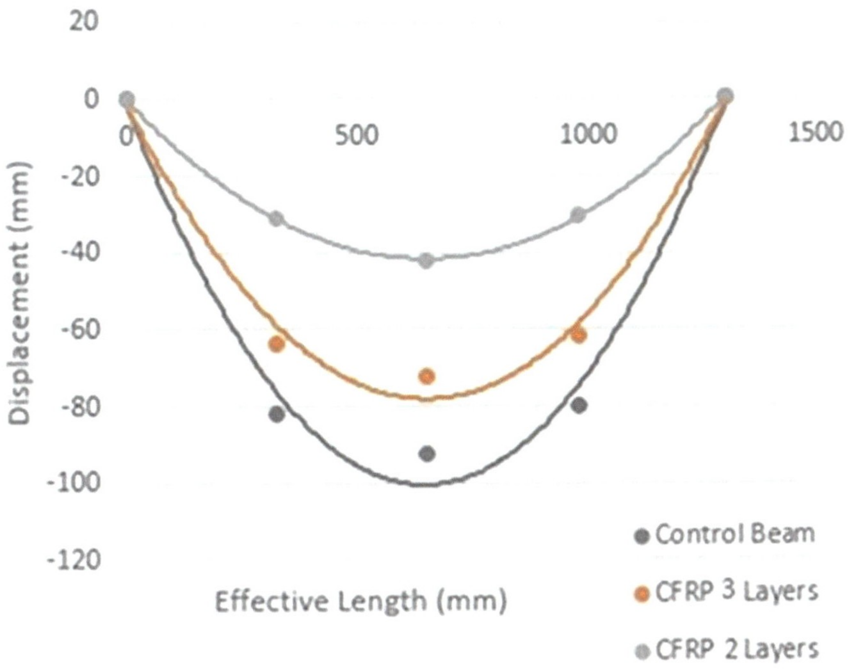

Fig. 2 Graph of displacement versus effective length

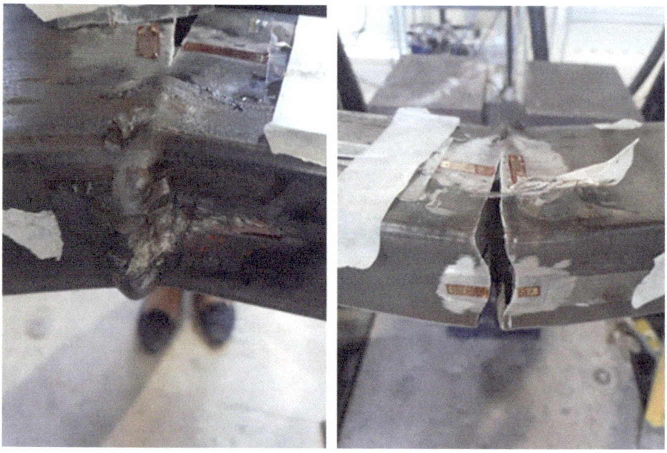

Fig. 3 Failure of the control beam: top side (left), bottom side (right)

Fig. 4 Failure mode of the beam B-3L

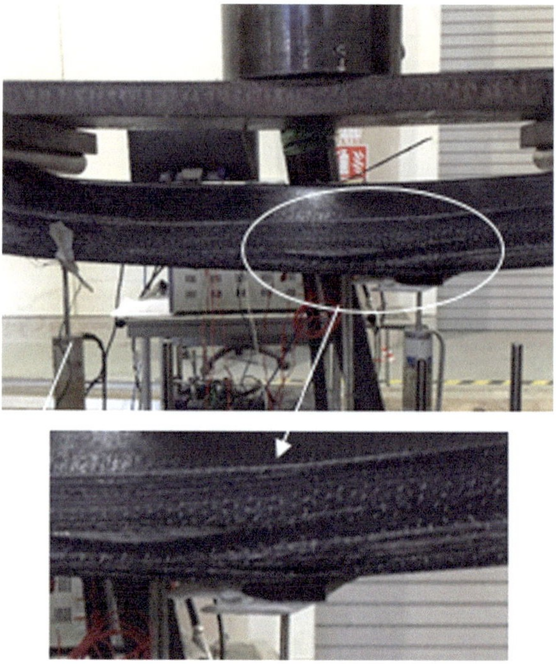

remain undamaged due to the low deformation capacity and consequently low level of stress concentration at the surface of the steel tubes. However, the control beam and the B-3L showed very ductile behaviour because of the yielding of steel tube prior to the concrete crush, as depicted in Figs. 3 and 4. Buckling of steel tube was observed at the top side of the control beam (B-C), whereas slippage of CFRP layers

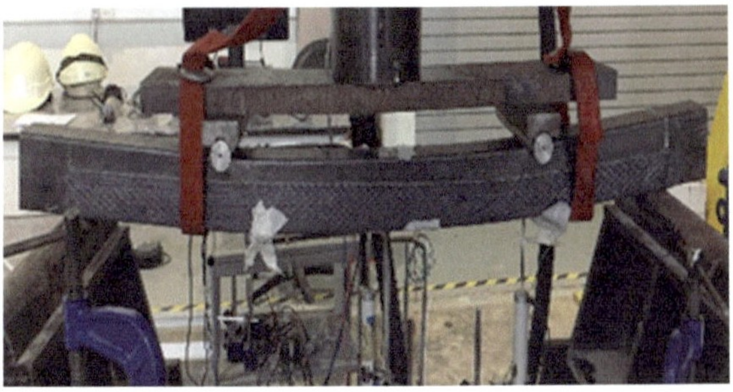

Fig. 5 Failure mode of the beam B-2L

occurred in the beam B-3L. There is no crack at the bottom and no buckling happened at the top of the beam as shown in Fig. 5.

4 Conclusions

In this study, structural behaviour of HPCFDST beams strengthened with CFRP sheets was investigated. High performance self-compacting concrete without coarse aggregates was used as a core between inner and outer steel tubes. The beams were experimented under pure bending tests. The results showed that the beams with CFRP layers exhibited 19% higher load carrying capacity compared to the control beam. Furthermore, it was shown that increasing the number of CFRP layers from two to three had little effect on the ultimate strength of the beams. The beam with three layers of CFRP showed a ductile behaviour due to great confinement provided by CFRP layers, whereas the beam with two layers of CFRP showed brittle behaviour due to crushing of concrete prior to the yielding of steel tubes. Slippage of CFRP layers occurred for the beam with three layers at the load of 64 kN.

Acknowledgements The authors wish to acknowledge the financial support received from the Fundamental Research Grant Scheme of Malaysia [grant number FRGS/1/2018/TK01/UTP/02/9].

References

1. Shanmugam NE, Lakshmi B, Uy B (2002) An analytical model for thin-walled steel box columns with concrete in-fill. Eng Struct 24:825–838
2. Ziyad AK, Gupta PK, Ahuja AK (2013) Parametric investigations on behaviour of square CFST columns. Int J Sci Eng Res 4(5):107–110
3. Kwan AKH, Dong CX, Ho JCM (2016) Axial and lateral stress-strain model for concrete filled steel tubes. J Constr Steel Res 122:421–433
4. Choi IR, Chung KS, Kim CS (2017) Experimental study on rectangular CFT columns with different steel grades and thicknesses. J Constr Steel Res 130:109–119

5. Chen J, Wang J, Xie F, Jin WL (2016) Behaviour of thin-walled dodecagonal section double skin concrete-filled steel tubes under bending. Thin-Walled Struct 98:293–300
6. Han LH, Huang H, Tao Z, Zhao XL (2006) Concrete-filled double skin tubular (CFDST) beam-column subjected to cyclic bending. Eng Struct 28:1698–1714
7. Uenaka K, Kitoh H (2011) Mechanical behavior of concrete filled double skin tubular circular deep beams. Thin-Walled Struct 49:256–263
8. Farajpourbonab E (2017) Effective parameters on the behaviour of CFDST columns. J Appl Eng Sci 418:99–108
9. Somayaji S (2004) Civil engineering materials. Prentice Hall, Michigan
10. Ahmed WA, Wan HWB, Azrul AM, Salam JH (2016) The enhanced performance of CFST beams using different strengthening schemes involving unidirectional CFRP sheets: an experimental study. Eng Struct 128:184–198
11. Ahmed WA, Wan HWB, Azrul AM, Salam JH (2017) Rehabilitation and strengthening of high-strength rectangular CFST beams using a partial wrapping scheme of CFRP sheets: experimental and numerical study. Thin-Walled Struct 114:80–91
12. Ahmed WA, Wan HWB, Azrul AM, Salam JH (2018) Flexural behaviour of CFST beams partially strengthened with unidirectional CFRP sheets: experimental and theoretical study. J Compos Constr 22(4)
13. Nikbakht E, Al-Fakih A, Hui CC, Jake LY, Mahzabin MS (2019) An experimental investigation on the shear and flexural behavior of steel reinforced HPSCC beams. Structures 19:286–295

Evaluating Passenger Load Factor of Public Bus Services in West Klang Valley

Shuhairy Norhisham, Muhammad Fadhlullah Abu Bakar, Zainul Hazwani Abd Samad, Agusril Syamsir, Nur'atiah Zaini, Herda Yati Katman and Mohd Zakwan Ramli

Abstract The Klang Valley Network Revamp was established in 2015 by Suruhanjaya Pengangkutan Awan Darat (SPAD) to improve the public transportation service quality in the Klang Valley. Upgrading the quality of bus services is one of the government's initiatives to improve public transportation in urban areas. This study aimed to evaluate the passenger load factor of the public bus service in three different areas of the Klang Valley. Three routes, each at five different areas in the west side of Klang Valley were selected, which covered Klang, Shah Alam, Puchong, Kelana Jaya, and Petaling Jaya, in order to identify the total passenger load at these selected areas. The results showed that most of the routes have the Quality of Service (QOS) A, and there were several aspects that should be improvised, such as the waiting time during peak hours and the various destinations in the area of service.

Keywords Public bus · Services quality · Passenger load · Transportation · Quality of service

1 Introduction

Buses are the most important public transport services in the Klang Valley. Buses have the most passengers among the public transport services and make approximately 600,000 trips each day [5]. The major bus company that operates in the urban areas of Klang Valley is the RapidKL service brand operated by Prasarana Malaysia. There were eight bus corridors that were revamped in December 2015 by the Land Public Transport Commission (SPAD) that plies the Klang Valley area, which used to be called the Klang Valley Network Revamp [9]. There are several attributes that affect the service performance of different modes of public transportation. Bus service

S. Norhisham (✉) · M. F. Abu Bakar · Z. H. Abd Samad · A. Syamsir · N. Zaini · H. Y. Katman · M. Z. Ramli
College of Engineering, Universiti Tenaga Nasional, 43000 Kajang, Selangor, Malaysia
e-mail: Shuhairy@uniten.edu.my

S. Norhisham · A. Syamsir
Institute of Energy Infrastructures (IEI), Universiti Tenaga Nasional, 43000 Kajang, Selangor, Malaysia

M. Awang and M. R. Meor M Fared (eds.), *ICACE 2019*,
Lecture Notes in Civil Engineering 59, https://doi.org/10.1007/978-981-15-1193-6_11

performance can be defined by attributes such as hour of service, service frequency, and passenger load factor [12, 15]. A better service is needed for existing buses and routes to be put in place to address under-served areas and neighbourhoods [9].

The Land Public Transport Commission of Malaysia had implemented a Bus Network Revamp (BNR) on 1 December 2015 [2]. Under the BNR implementation, existing bus corridors were reorganised into smaller corridors, namely the Jalan Ipoh Corridor, Jalan Pahang Corridor, Ampang Corridor, Cheras Corridor, Sungai Besi Corridor, Klang Lama Corridor, Lebuhraya Persekutuan Corridor, and Damansara Corridor. The Klang Valley area lies in the Klang Lama Corridor, Lebuhraya Persekutuan Corridor, and Damansara Corridor, which are serviced by RapidKL, Metrobus, Cityliner, Seranas, and Causeway Link [6, 20].

In the towns of Klang, Shah Alam, Petaling Jaya, Subang Jaya, and Puchong, the main public transportation system is the bus service [10]. However, several news reports had been published pertaining to poor bus services in the Klang Valley [3, 19]. Therefore, there is a need to study the current state of those five particular areas of public buses in order to provide a thorough glimpse of their service ratings and performances [9, 13]. As for the bus services, most of the people who use Rapid KL have resigned themselves to the fact that most of the time it seems like they do not have a schedule or a timetable that they can stick with regularly [8, 16]. In addition, there are numerous problems plaguing the passengers in the Klang Valley, such as long waiting time (Service Frequency) [11], overcrowding, especially in peak hours (Passenger Load Factor) [7], and poor management of bus service hours, which lead to delays (Hour of Service) [1, 4].

The question that should be addressed regarding this issue is why the bus services in the Klang Valley area do not have a good performance; on the other hand, the quality of service, particularly in terms of service frequency, hours of service and passenger load factor are at a satisfactory level [14, 18]. In this study, the areas covered are the west side areas of Klang Valley, which are the towns of Klang, Shah Alam, Petaling Jaya, Subang Jaya, and Puchong, wherein there has been no comprehensive study or paper published on the public bus services. Hence, in this study, several attributes relating to the service frequency of public buses were identified and an evaluation of passenger load factor and the hour of service of public buses were determined.

2 Methodology

The data collection activities were conducted at selected routes of RapidKL Bus. For a RapidKL Bus, the total capacity is 65 passengers, consisting of 25 seated and 40 standing passengers. Meanwhile, the occupancy of bus was counted by recording the in and out passengers at every bus stop, for at least three routes in each five different areas in the west side of Klang Valley. The routes recording the in and out passengers were Klang (Route KLG1, KLG2, and 704), Shah Alam (Route T757, T758, and SA01), Subang Jaya (Route T776, T777, and T778), Petaling Jaya (Route PJ02,

PJ03 and PJ04), and Puchong (Route T600, T604, and T605). The data collection was done in different time frames identified as below:

- Peak Hours in the Morning (7 am–9 am)
- Non-Peak Hours of the Day (11 am–2 pm)
- Peak Hours in the Evening (4 pm–7 pm)
- Non-Peak Hours of the Night (8 pm–10 pm).

The data collection included weekdays and weekends. The total journeys for the data collection were four journeys for weekdays and four journeys for weekends, for each respective route in one area only. To sum up, the total journeys for one area was 24 journeys, while the total journeys for the total five different areas were 120 journeys altogether. The conditions inside the buses are shown in Fig. 1.

As per the journey, the average passenger load was counted and the number was divided to the number of seats to become the load factor. For example, if the average passenger load of a bus was about 15 passengers and the number of seats was 25, so the load factor was 0.60. Table 1 shows the quality of services for load factor with QOS B for 0.60 load factor.

The analysis gave a QOS for every three routes in five different areas of the west side of the Klang Valley for the load factor. However, the consideration of QOS for every journey, on weekends and weekdays and peak hours or non-peak hours should be differentiated. The results are presented in a graph to compare all the QOSs in every route and a summary was made for the QOS for all three routes in each of those five areas. Passenger load was calculated by dividing the total passenger by the total seats provided in each bus as shown in Eq. 1.

$$\text{Passenger Load} = \text{Load Factor} = \text{No of Passengers/No of Seats} \quad (1)$$

3 Results and Analysis

The data were collected for each selected route for each location. All the data were summarised in Fig. 2. The three routes of each location from Klang (Route KLG1, KLG2 and 704), Shah Alam (Route T757, T758 and SA01), Subang Jaya (Route T776, T777 and T778), Petaling Jaya (Route PJ02, PJ03 and PJ04), and Puchong (Route T600, T604 and T605) are presented as Routes A, B, and C.

From the graph in Fig. 2, it can be seen that the lowest passenger load factor at Puchong was for route T605 (C) compared to the routes T600 (A) and T604 (B). The reason could possibly be because the places covered by route T605 (C) were not interesting as it mostly covered residential areas and the route was also quite short compared to the other two routes. As for the route T600 (A), it did cover places like shopping malls, which were convenient to passengers. Meanwhile, the load factor at the Petaling Jaya area showed that the lowest passenger load factor was for route PJ03 (B) compared to routes PJ02 (A) and PJ04 (C). On the other hand, route PJ02 (A)

Fig. 1 Data collection of the selected routes

had the highest passenger load factor among the three routes, and the route mostly covered office areas where the public went to deal with business or personal matters. Thus, route PJ02 (A) had a high passenger load factor compared to routes PJ03 (B) and PJ04 (C).

Comparison of passenger load factors could be made since these studies have covered five different areas. To illustrate, a comparison can be made by riding all RapidKL buses that covered the three routes for Puchong and Subang Jaya. On the other hand, for Shah Alam, two RapidKL buses and one SMART Selangor bus were

Table 1 Quality of services (QOS) for load factor (*Source* Transit capacity and quality of service manual [17])

Quality of service	Load passenger (passenger/seat)	Comments
A	0.00–0.50	No passenger needs to sit next to another
B	0.51–0.75	Passenger can choose where to sit
C	0.76–1.00	All passengers can seat
D	1.01–1.25	Comfortable standee load for design
E	1.26–1.50	Maximum schedule load
F	>1.50	Crush load

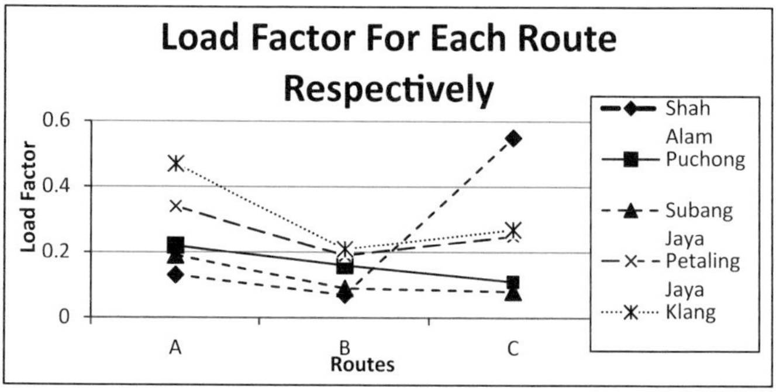

Fig. 2 The graph of load factor for routes at each location

taken in order to see the differences between these two operators, and it can be seen that the load factor for the SMART Selangor bus that covered Shah Alam was quite high compared to RapidKL buses which fell into QOS B. Apart from that, for Petaling Jaya, the PJ City Bus was chosen since it is a free bus service, and also, to see the consistency of this operator. As it can be seen from the comparison, there was only a slight difference and the three routes still fall into QOS A. In this study, comparison of the average of load factors on two different situations was also identified. Figure 3 shows the average load factor at the five selected areas. The graph shows that the lowest average load factor was at Subang Jaya on a weekday, which was 0.11, and the highest average of load factor was Klang on a weekend.

Passenger load is the factor between the passengers to the total number of seats on the bus, where the load factor can be obtained in order to determine the QOS. For the present study in the Klang Valley, the area is very wide and a few different operators including RapidKL for Puchong, Subang Jaya, and another two routes for

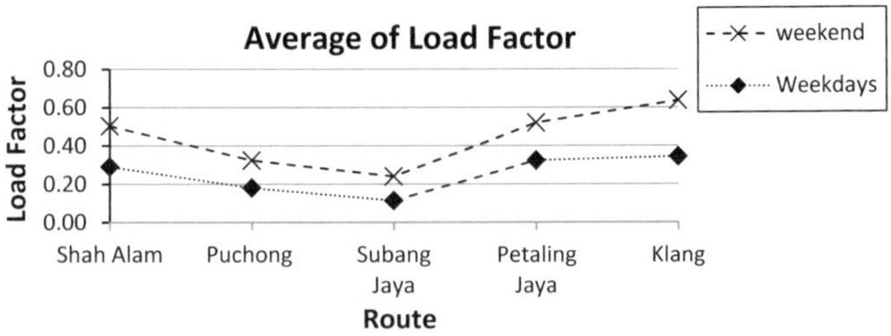

Fig. 3 The graph of average load factor on weekdays and weekends

Shah Alam were studied. Apart from that, other operators were the PJ City Bus which only covered Petaling Jaya, while one route for Shah Alam and two routes for Klang were covered by the SMART Selangor free bus service. Comparison can be made in the study as the data have covered five different areas in Klang Valley, where the comparison can be made by riding all RapidKL buses that covered three routes for Puchong and Subang Jaya. On the other hand, for Shah Alam, two RapidKL buses and one SMART Selangor bus were taken in order to see the difference between these two operators and it can be seen that the load factor for SMART Selangor bus that covered Shah Alam was quite high compared to RapidKL buses which fell into QOS B.

Apart from that, for Petaling Jaya, the PJ City Bus was chosen since it is a free bus service, and also, to see the consistency of this operator. As can be seen from the comparison, there was only a slight difference among these three routes which all fell into QOS A for weekdays and weekends. Besides that, SMART Selangor and Seranas buses were chosen for the Klang area in order to see the differences and the efficiency of the new free bus service provided by the Selangor State Government where the service was quite new compared to the Seranas bus which was really old in providing bus services and it was not free. So far, for the passenger load factor, all three routes for each of the five areas fell into QOS A where no passengers need to sit next to another, except for the SA01 (A) route for Shah Alam, which fell into QOS B where the passenger can choose where to sit, according to the Transit Capacity and Quality of Service Manual for Load Factor.

4 Conclusion

This study was carried out to evaluate the passenger load factor of the public bus service in five different areas. The results showed that most of the routes in the selected areas had an average passenger load factor of less than 0.50, which fell into the A quality of service. There was only one route in Shah Alam, SA01 (C), which had an average passenger load factor between 0.50 and 1.0, which fell into the B service quality. Besides that, weekdays had a lower average of passenger load factor compared to weekends. The result indicates that the average of passenger load factor of public bus service would be determined based on certain routes and specific time. There are certain routes and specific times which have high average of passenger load factors, such as routes for residential, university, and shopping complex areas. The outcome of this study is very important to collect all necessary data and information to improve public transport services in the Klang Valley. There are several routes that could be added in the selected location in future, such as in Petaling Jaya, Shah Alam, and Puchong, especially in peak hour period. These findings could also be as key indicators for the effectiveness of the prior government's policy. Further studies in this area with a more comprehensive scope and coverage are highly suggested.

Acknowledgements The authors would like to acknowledge the Internal Research Grant provided by Universiti Tenaga Nasional (Project No: J510050878) for this research.

References

1. Abdullah AA (2013) Rapidkl bus service in City Center, Kuala Lumpur, Malaysia: an epitome of good service? Int J Acad Res Bus Soc Sci 3(4):333–349
2. Amiril A, Nawawi AH, Takim R, Latif SNFA (2014) Transportation infrastructure project sustainability factors and performance. Procedia Soc Behav Sci 153:90–98. https://doi.org/10.1016/j.sbspro.2014.10.044
3. Bachok S, Osman MM, Ponrahono Z (2014) Passenger's aspiration towards sustainable public transportation system: Kerian District, Perak, Malaysia. Procedia Soc Behav Sci 153:553–556. https://doi.org/10.1016/j.sbspro.2014.10.088
4. Chang SKJ, Hsu CL (2001) Modeling passenger waiting time for intermodal transit stations. J Transp Res Board (1753):69–75. https://doi.org/10.3141/1753-09
5. Chuen CO, Karim RM, Yusoff S (2014) Mode choice between private and public transport in Klang Valley, Malaysia. Sci World J 2014:14
6. Eboli L, Forciniti C, Mazzulla G (2014) Service coverage factors affecting bus transit system availability. Procedia Soc Behav Sci 111:984–993. https://doi.org/10.1016/j.sbspro.2014.01.133
7. Hazril MI, Sabri SMI, Tajedi NAA (2013) A study on previous and ongoing research regarding public transportation in Malaysia. In: Proceedings of Malaysian technical universities conference on engineering and technology (MUCET), Kuantan, Pahang, Malaysia, 3–4 Dec
8. Leong LV, Ibrahim WHW, Sadullah AFM (2006) Passenger car equivalents and saturation flow rates for through vehicles at signalized intersections in Malaysia. In: ARRB conference—research into practice 2006, Canberra, Australia
9. PEMANDU (2012) Government transformation programme the roadmap 2, vol 0, Prime Minister's Department
10. Ponrahono Z, Bachok S, Osman MM, Ibrahim M, Abdullah MF, Abdullah A (2015) Assessing the urban and rural stage bus services disparities in Peninsula Malaysia. J Malays Inst Plan 13(2015):65–84
11. Redman L, Friman M, Gärling T, Hartig T (2013) Quality attributes of public transport that attract car users: A research review. Transp Policy 25:119–127. https://doi.org/10.1016/j.tranpol.2012.11.005
12. Rohani MM, Wijeyesekera DC, Karim ATA (2013) Bus operation, quality service and the role of bus provider and driver. Procedia Eng 53:167–178. https://doi.org/10.1016/j.proeng.2013.02.022

13. Sabra WA (2003) A comparison between highway capacity manual and critical lane volume analysis methodologies for traffic impact study applications, 1–7
14. Shariff NM (2012) Private vehicle ownership and transportation planning in Malaysia. In: International conference on traffic And transportation engineering, vol 26, pp 64–68
15. Suwardo W, Napiah M, Kamaruddin I (2008) Punctuality and expected waiting time of stage buses in mixed traffic. Jurnal Transportasi 8(3):213–226
16. Suwardo, NMB, Kamaruddin IB (2009) On-time performance and service regularity of stage buses in mixed traffic. World Acad Sci Eng Technol 3(7):1137–1144
17. Transportation Research Board (2013) TCRP report 100: transit capacity and quality of service manual, 3rd edn. Transit Cooperative Research Program, The Federal Transit Administration, Washington, DC
18. Transportation Research Board (TRB) (1999) A handbook for measuring customer satisfaction and service quality. TCRP report 47
19. Vien LL, Wan A, Wan H (2006) Passenger car equivalents and saturation flow rates for through vehicles at signalized intersections in Malaysia, 1–9
20. Yaakub N, Napiah M (2011) Public transport: punctuality In-dex for bus operation. World Acad Sci Eng Technol 60(12):12–23

Latent and Non-latent Building Defects: A Literature Review

Roslan B. Talib and Mohd Zailan B. Sulieman

Abstract The alleged building defects are adversely affected starting from the design part as the first defect possibly to have emerged. So far, there is no genuine comprehensive guide that has been carefully gathered in an organized manner to comparatively table out the design part focusing on building defects: latent or non. As usual, the defects are expected to be existing right after the building's rigorous examination, especially during the building inspection period. It is invariably the ambitious target of this empirical research to accumulate all the essential parts and key pieces in an organized way, which is caused and found within the identified twelve equivalent categories such as on building construction components that produce defects effect; indisputably seen or cleverly hidden or covered up. It is the key target of the mission, anticipating to produce referred guidelines so that the people involved within the building industry can look upon. The research intention, somehow, is to look for the gap which, so far, nobody in the industry has attended to. Once the gap has been positively identified, the proper and systematic cataloging system can be undoubtedly produced and the fewer defects, latent or non-latent, can be squeezed out and the construction process will be obtained at its excellent quality by eliminating building defects. The paper concluded five main factors that greatly contributed to cause the defects which are design factors, lacking referred data, water seepage, dampness, and defects caused by fire.

Keywords Building defects · Construction defects · Latent defects · Non-latent defects · Unseen-defects

1 Introduction on Defects

Built environment is a complex multidisciplinary field which propels efficient and effective research that goes beyond the traditional approach [1–3]. As part of the built environment scope, the present construction industry sector concomitantly within

R. B. Talib (✉) · M. Z. B. Sulieman
School of Housing Building and Planning, Universiti Sains Malaysia (USM), 11800 George Town, Pulau Pinang, Malaysia
e-mail: roslantalib@student.usm.my

M. Awang and M. R. Meor M Fared (eds.), *ICACE 2019*,
Lecture Notes in Civil Engineering 59, https://doi.org/10.1007/978-981-15-1193-6_12

the Industry Revolution 4.0 era steps into the developed world and is getting more industrialized, getting more advance and emergent day by day, undoubtedly including Malaysia. Despite the development, construction industry is confronting one fundamental problem which is building defects. Project Managers, Building Maintenance Managers, Building Contractors, and Design Architects or even Interior Architects are consistently striving to overcome the challenge of defects in buildings, but it is difficult to deal with it completely.

To start with, according to Jaspal et al. [4], a defect is invariably a building flaw or design mistake that progressively reduces the economic value of the building and it causes not an immaculate condition to the buildings. A construction defect can arise due to many factors like poor workmanship or the use of inferior materials. Building defects do not appear to be minimized despite recent advancements in building technology [4]. Some common defects caused by agents such as atmospheric pollution, poor workmanship, or the use of inferior materials and climatic conditions are more frequently found. Defective building construction does not merely contribute to the rise of the final cost of the product, but also to the cost of maintenance, which can be substantial.

Not only the defects are inadvertently caused but also can be seen by physically visual above-grade condition; however, to make it latent, building or construction defects have also occurred below ground as well. Given building foundation as one of the examples, defective construction includes activities like inadequate compaction of the soil not done to specifications which can lead to ground subsidence and an eventual early deterioration of foundations. This may lead to the catastrophic failure of a structure, and this is very dangerous to the austere exterior as well as the interior of the building.

Conditions under which building construction takes place are often far from ideal with the focus mainly being on speedy delivery. Defects resulting from inaccurate construction method can be avoided by ensuring that proper inspection mechanisms are in place. The comprehensive understanding of building defects and their apparent causes are essential for better performance of any building. In general, building defects fall into two categories: first, the defects that affect the performance and function of the building and second, the defects that affect the external and internal visual of the building.

2 Research Intent

The main objective of this research is the preliminary examination of defects and to lessen the cost of construction, as well as improving the quality of materials used in the construction. The occurrence of defects in construction will lead to a lower satisfaction of the customers as well as a decrease in the progress reputation in the building market: private or public sectors. Project owners have the right to own quality buildings with correct construction details especially for the Government who own public buildings.

Simply, the most primary reason is shoddy workmanship as they do not follow the instructions which were given in the specifications; thus, they are responsible for the occurrence of defects. A comprehensive inspection of work is necessary for a building at a time interval throughout the life of a building. The lack of maintenance or incorrect maintenance will transcend to reduce the life of the building and reduces the effective life of the materials far lower than it should be achieved [5]. This succeeding stage of the research is to concentrate on defects that occur in construction projects after the handover of the project. Defect analysis is not a safety task because the life of the building's safety normally depends on the strength and durability of the components of building and if the components are defect-free; ultimately, the owner will get the best service life and safety.

Subhi et al. [6], Glover [7], and the National Building Agency [8] stated that research shows that due to the construction defects, the cost of construction has been increased and the life of the internal structures of the buildings are typically decreased. The practical reason for the defects commonly found everywhere as mentioned before, i.e., due to poor workmanship, improper design, an essential quality of materials, lack of supervision, wrong construction methods, or lack of comprehensive inspection after construction, etc. [6–8].

After doing an intensive review on the rebellious subject, I had inevitably come up with the graphic (Fig. 1) summarizing how to locate the unbridgeable gap to be anxiously studied. After that, the next stage is to come up with an appropriate theory to correctly solve the current scenario. The significant categories are somehow related to each other. But somehow it was improperly unmeasured to reluctantly produce the systematic identified defect issues or the GAP which can be used as a stringent guideline derived from the peculiar four phase as indicated; thus, hoping to satisfactorily solve the subject matter problems once and for all.

At this point, the loophole allegedly happened when there were no critical linkages among the four distinct phases of the construction period. It started from the flimsy preconstruction period especially during the design process. Then during the construction phase and during the post-construction period alleged defects were inadvertently found, regardless latent or non-latent. These fundamental defects must be

Fig. 1 Four phase defects components in building construction categories overly with the research GAP

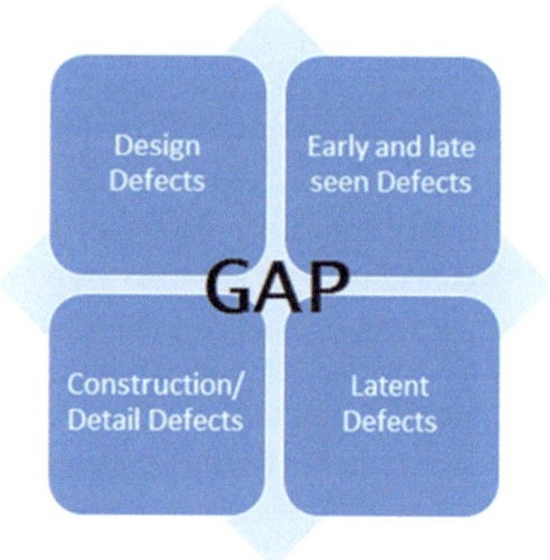

documented properly and systematically in a correct and proper order to intentionally build in a reference guide for the related defects to be used appropriately to rectify the ethical problem at any stage period of the process.

This conceptual paper is trying to conceptualize how the accumulation of all the building defects, either latent defects or non-latent defects, can help to adjust the existing theories from related or other fields, then it can help modify the framework onto the new field in order to construct our own theoretical framework. The aim somehow is also to validate and modify the existing concepts according to the conclusion results of the literature review. This can then even lead to a revision of the concepts and theories or even developing new models that will be tested for the next stage.

3 Latent Defects

The other problem is that we cannot see and find out what the defects of the buildings are. Some of these building defects cannot be searched diligently or viewed favorably. Moreover, during the handover time, it is found that the building is in good condition, but upon reaching the expired level, it is found that the building was not. Researches on this type of building defect address this as latent defect. This latent defect constitutes a fundamental problem in the construction industry as it is indistinguishable and undetected. In the law of the sale of property (both real estate and personal property) a latent defect is a fault in the property that could not have been discovered by a reasonably thorough inspection before the handover period to the tenant or client. Identical concept is applied to new Government constructed buildings as well.

Furthermore, a latent defect is a flaw in building property you cannot easily spot with a superficial inspection and that the building owner does not know about its existence [9, 10]. This contrasts with a patent defect, which is a flaw that is relatively obvious that you can usually spot easily. For example, if you represent the owner of a building and notice a deep crack in the wall or a door with a hole, then these represent examples of patent defects because you can recognize them quite easily. However, latent defects could be rooms painted with lead-based paint, for example, or a mould problem under the carpeting. These defects are not easily discoverable in the absence of an inspection, and it is possible that the owner does not know about them at all (Quickbooks.intuit.com/ca (2019)) [11].

Chong et al. [12] stated that it was impossible to identify all types of latent defects without a good information source to rely upon. Designers frequently rely on their own experience, some employ consultants, and a few have some sort of database of past defects to guide them. But experience and research have showed that many latent defects were repeated in many of the buildings and designers were unaware of them. We must accept that it is impossible to eliminate all latent defects in a building. However, it is possible to eliminate most latent defects through better design only if the designers know about them. Designers are encouraged to manage a database whereby latent defects could be traced and mistakes could be avoided in their following designs. Such a database can be developed with existing standards and the information must be properly gathered systematically, especially from property managers [12].

4 Defects That Cause Fire

Looking from divergent perspective, fires started in buildings are usually caused by careless use of open fire (kitchen burners or even from cigarette butt), the failure of indoor installation of building's mechanical or electrical systems, or by lightning. These defects related category, however, did not include fire which was created after the gas explosion, for example. Interestingly, structural collapse leading to building explosion or fire leading to collapse or damage to the construction are not among the causes leading to the identified building defects.

Other than the abovementioned reasons, there are many causes for fires to occur, especially in public buildings. Among the most common reasons are the representatives of inadequate design of fire protection. This category can be accurately classified into the design failure that can contribute to building defects. McDonald [13] simply explained the concerned terms; first, making adequate provisions to prevent the structure from collapsing for a predetermined time required for occupants to allow them to vacate the building; second, providing suitable means of escape; third, having suitable methods to prevent the fire from spreading by creating fire-resistant compartments to contain the fire. Subsequently, poor building management by contributing in allowing fire hazards to accumulate (such as rubbish, flammable furnishings, and materials, etc.) [13]. Furthermore, lack of maintenance: faulty electrical installations or malfunctioning sprinklers, alarms, etc. Lastly, careless acts such as work using flames, or smoking, and lack of staff training for fire emergencies as well as human error/negligence are considered to contribute to buildings defects term.

McDonald [13] again, mentioned that fire affects the number of lives lost and this is caused by the smoke-induced asphyxia, collapse of the structural elements, and victims jumping out of windows. First, the fire destruction of the building is due to its fabric component elements and the contents of the flames. The fire will utilize every available material as fuel: timber structure, furnishings, fabrics, finishes, etc. and consumes it in the process. Second, the destruction, alteration, and damage caused to the building elements by the excessive temperatures will cause them to expand, crack, vitrify, plastify, melt, etc. [13]. Then, the damage caused to the fabric and contents by the enormous quantities of water used to extinguish the fire. Last but not the least, the damage caused by the mechanical collapse of structural elements of the buildings hugely impacted on the interior part of the buildings. So, building defects which have an impact on the creation of fire within the buildings are among the most significant categories to investigate.

5 More Review on Literature

The other part of building construction is to deal appropriately with building maintenance when the buildings are ready to be occupied. However, building maintenance remains an expansive item which requires careful attention from building owners, occupiers, contractors, surveyors, architects, and in fact, every active member of the construction industry. The National Building Agency (NBA) headquarters in Dublin, Ireland, stated that the accurate diagnosis of building defects associated with the correct remedial action, serve as merely an economic basis for a successful building maintenance programme [8]. Defects diagnosis merely remains as a logical way of proceeding from the evidence to the cause of a defect, after which remedies can be prescribed. It is critical not to have preconceived ideas on causes, nor to jump to conclusions, otherwise, an incorrect remedy may be put in hand, perhaps more expansive than necessary [14–16].

Mahli et al. [17] studied that the critical age of a building condition is within the range of 11–20 years old where the number of building defects keep on increasing. This is supported by their finding that two schools were found in dilapidated conditions which were aged more than 20 years, within their 24 schools being studied in the State of Sarawak [17].

With poorly repaired and cracked defects opened up again (due to the continuing movement of the building wall), the interior space is seen physically greatly impacted. It is necessary in the client's best interests to refer a problem to a contractor rather than to a professionally independent structural engineer. Glover [7] suggested that the first matter to be carefully considered would be the available evidence of actual past and the considerable damage to the structure under review, with regards to age and all circumstances. Initially, any aboveground damages would need to be assessed, for example, cracks in walls (inside and outside) and evidence of cognitive distortion to windows and doors, reasonably not too deep soil foundation, sloping floors, and irregular ceilings features [7].

Mahli et al. [17] again found that the building defects happened mainly on the ground level with the most significant number of defect components found on the walls. 16% of defects are wall cracks from 11 common defects and most of all the highest score of defects happened to buildings aged 11–20 years [17]. However, Harris [18] viewed and stressed out that unlike paintings or sculptures, people occupy buildings, which makes those people vulnerable to the prospect that the building will collapse, or that they may be trapped inside the building in a fire or trampled in their effort to escape. There is no law or requirement that people must enter a building regardless of the possible risk. Despite our apparently fervent desire to allow persons to enter or occupy old or historic buildings, there is no absolute requirement for them to do so [18].

Cook and Hinks [19] indicated that many materials used in construction are porous. Ceramic and masonry components retain various degrees of porosity depending on their composition and manufacturing, where cycles of wetting and drying can make the material lose any water-soluble compounds present within it. This will expose a greater surface area to further effects, increasing the risk of immune deficiency [19]. The moisture cycles will also cause movement of the material, which will adversely affect durability process, thus, enhancing the building defects. Harris [18] clearly stated that people did not have certain knowledge and they cannot gain complete control over the progress of deterioration. As buildings age, the lines of stiffness shift, the load-carrying distribution within the structure changes, and at the macro-scale, no-load bearing elements begin carrying the load [18].

Harris [18] subsequently mentioned that the current state of design practice de-emphasizes responsibility for the future of the building and there is a disjunction on the part of the design totality from the world of cause and effect. If the designer calls for a lime-stone façade in an acidic environment, the consequences are as predictable as when the preservationist restores the same limestone item in a similar acidic environment. Whereas the preservationist is expected to recognize the effects of acids on the ionic salts, for reasons related to the emphasis on novelty and entertainment, the architect can consider the limestone art and decline responsibility for its otherwise predictable dissolution [18]. Designers and educators of designers must ensure that buildings look good and they perform longer than it is required for the pretty initial photo shoot. It is not the image just enough to satisfy our definition of design.

However, Cook and Hinks [19] indicated that where water penetrates organic material, it can generate suitable conditions for the growth of fungi and moulds. These can undoubtedly break down the structure of the material, causing discoloration and substantial loss of strength. The moisture content to support fungal attack is frequently assumed to be 20%, although, once established, some rots can affect material with lower moisture contents. In the case of timber, high moisture contents are also associated with reductions in strength and dimensional instabilities [7, 19], where the timber is subjected to drying and wetting cycles, the dimensional changes will stress the already weakened timber and may hasten further deterioration.

Mills et al. [20] simply mentioned that defects have constituted a recognized part of the building process. The most frequent defects were the roof leaking, albeit based on a limited number of samples from the HGF database. The most expansive water ingress claims occurred in the leaking windows and external water penetration categories [20].

Lastly, Talib et al. [5] concluded that all the identified internal defects are typically caused by one main treat which is water. They stated that there were five categories that are derived from the five main internal defects which are dampness, cracks, detach, wear and tear, and biologically caused defect category [5]. With the condition of shimmering haze that has been getting worse during the time of the article being written, it is important to us to have an ideal internal detail design with minimum defects as we may have to spend more time inside of a building than the outside.

6 Conclusion

Dampness and unmistakable cracks represent common manifestation of possible defects. Interior dampness can, however, frequently lead to cracks, undoubtedly making it difficult to correctly determine the root cause and the appropriate action. Talib et al. [5] succinctly summarized that building interior related defects can be conveniently divided into five (5) significant categories which are invariably dampness, visible cracks, detach, wear and tear, and biological category [5].

The entire buildings are subjected to various forms of defects, failures, deterioration, and variations [21]. The literature has explored several building defects and contributing factors which can be associated with the major theme of this research. Bakri and Mydin [21] stressed that it is significant to appraise each defect and ultimate failure in every key part of the building and assuredly find out the primary causes of each individual defect and failure. Under those circumstances, the remedy procedure should be done correctly. The contributing factors to these fundamental defects and failures must be investigated intensely. Once the possible causes of the defects and failures were found out, it is imperative to distinguish how to keep away from it in the future and reduce the effect to the minimum stage.

Having stated on the above matter, the graphic (Fig. 2) can be properly referred as a published summary to the book-related reviews invariably done at this qualifying stage. There are twelve optimal levels of distinct categories that have been pointed out, showed and assuredly found in the scholarly journals, books, or online data related to specific hidden defects or covered up fundamental defects in amicable relation to the building construction part. Even though defects can be undoubtedly related to the austere exterior of the notable buildings from the roof, building skins, or even the foundation part, however, it is the interior part of the buildings indisputably affected the most.

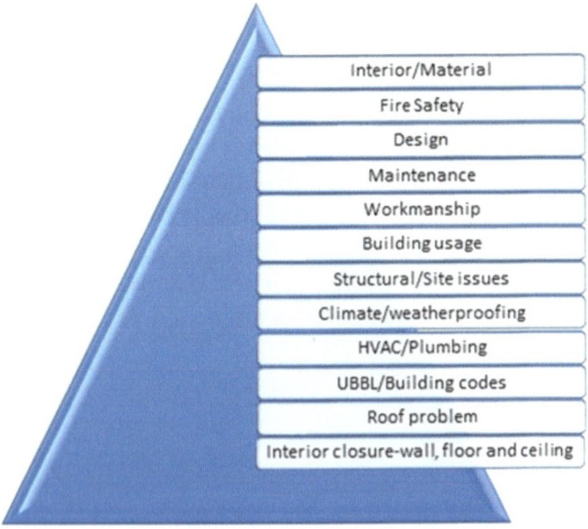

Fig. 2 Graphic summarization of 12 levels on literature research depicting on latent defects in the construction industry

The last summary can be written as promptly follows: (1) So far, no systematic crucial central documents on defects, especially on latent defects, have been produced as reference document to be used by designers, project managers, or building managers. (2) Factual data will be popularly used for this empirical research, consequently making this investigative report more comprehensive. Ready data are being accurately identified from the local contractor who specializes solely on the rectification of the building defects works. (3) Treat fire as the main contributor to latent defects as when the building got burned and destroyed, unknown real legitimate cause being positively identified but at the end, it is always due to the latent defects. (4) Water is traditionally the most top contributor for the fundamental defects: latent or non-latent, and extensive research must properly focus on this key issue, especially for civic buildings located within the tropical regions with more accumulated rain-fall compared to other non-tropical territories [7].

Acknowledgements We extend our gratitude to Dr. Linariza Haron of SHBP-USM for her contribution to this study as well as those in the Dean and Deputy Dean Office of Research and Administrate Office of SHBP-USM, USM's Hamzah Sendut Library 1 and IPS office, USM. Puan Siti of JTSB office in K Batas, SPU, P Pinang.

References

1. Abu Bakar AH, Roosli R, Kamal EM, Ahmad ZZ (2017) A complete guide to academic research in built environment. USM Publisher
2. Denscombe M (2011) The good research guide for small-scale social research projects, 4th edn. Open University Press, Buckingham
3. Lester JD, Lester JD Jr (2011) Principles of writing research papers, 3rd edn. Penguin Academics, Arizona, USA
4. Jaspal S, Kamaldeep K (2018) Defects in buildings causes and remedies. NBM&CW. NBM Media Pt. Ltd. https://www.nbmcw.com/tech-articles/tall-construction/29714-defects-in-buildings-causes-and-remedies.html. Accessed 19 Dec 2018
5. Talib R, Ahmad AG, Sulieman MZ (2015) Factor affecting typical interior defects (water based) found in Malaysian buildings. With real selected case samples. Int J Emerg Trends Eng Res (WARSE/IJETER) 3(8):55–64
6. Subhi M, Dinesh S, Resmi R (2017) A review on construction defects. Int J Inform Futur Res 4(8):7079–7084
7. Glover P (2003) Building surveys, 5th edn. Butterworth Heinemann
8. The National Building Agency (1979) Common building defects. NBA's Housing Assc. Maintenance training course. The Construction Press, Lancaster
9. Practical Law (2019) Glossary—latent defects. Thomson-Reuters. https://uk.practicallaw.thomsonreuters.com. Accessed 2 Jan 2019
10. Gauthier A (2018) Latent defects: the burden of proving hidden defects. https://www.alepin.com/en/legal-advice/351-latent-defects-the-burden-of-proving-hidden-defects.html. Accessed 10 Dec 2018
11. Attov A (2019) What is a latent defect? Intuit Inc. https://quickbooks.intuit.com/ca/resources/business-law/latent-defect-definition-e. Accessed 15 Dec 2018
12. Chong WK, Low SP (2006) Latent building defects: causes and design strategies to prevent them. J Perform Cons Facil/ASCE 20(3):213–221

13. McDonald R (2003) Introduction to natural and man-made disasters and their effects on buildings. Architectural Press
14. The Whitestone building maintenance and repair cost reference 2007–2008. In: 12 annual ed. Whitestone research
15. Kubba SAA (20018) Architectural forensics. McGraw Hill, New York
16. Graham D (2014) Assessing construction defects in high end fit-out construction. https://www.theexpertinstitute.com/assessing-construction-defects-high-end-fit-construction/. Accessed 15 Dec 2018
17. Mahli M, Che-Ani AI, Yahaya H, Tawil NM, Othuman Mydin MA (2014) School building defects pattern. In: Building surveying, facilities management and engineering conference (BSFMEC2014), pp 1–5. EDP Sciences. 20141501007
18. Harris YS (2001) Building pathology—deterioration, diagnostics and intervention. Wiley, New York
19. Cook GK, Hinks AJ (1992) Appraising building defects: perspectives on stability and hygrothermal performance. Longman Scientific & Technical
20. Mills A, Love PED, Williams P (2009) Defects costs in residential construction. J Cons Eng Manag 135(1):12–16 (@ASCE, 0733-9364)
21. Bakri NNO, Othuman Mydin MA (2013) General building defects: causes, symptoms and remedial work. Eur J Tech Des 3(1):4–17 (Academic Publishing House Researcher)

Lighting for Heritage Building: A Case Study of the Lighting Design Applied on St. George's Church in George Town, Penang Island

Alaa Abdalla Saeid Ali, Safial Aqbar Zakaria, Arran Chang Kien Guan and Ch'ng Jack Shun

Abstract Light is one of the critical aspects of architecture, yet it is one of the understated elements in our daily lives. Light has a significant spot to form events, activities, and memories. Scientific studies have shown that appropriate and proper lighting does not only affect human health but daily human moods as well. Nowadays, lighting fixtures play essential roles in architecture and markets. A proper lighting fixture helps to highlight the structure, textures, and form of the shape of a building. However, the designers tend to focus more on the art aesthetics rather than the conservation ethics, such as the colorful design and the attractive lighting fixtures on the building, while the prime concern should be on realizing the impact of the lighting fixture toward the environment. The objectives of this study are to highlight the issues of the lighting system of the heritage building and suggest some recommendations to meet the requirements for the heritage building for better lighting design. The data for this research was collected using the quantitative method. Thus, as a result, it is found out that lighting design highlights the historical building, and therefore, attracts tourists, which benefit the economy. However, improper lighting installation leads to a negative impact on the society and human health.

Keywords Historical building · Lighting design · Conservation · Preservation · Art and history · Light fixtures

1 Introduction

The design of heritage buildings in Malaysia has its uniqueness and meanings in every particular place, especially in Penang Island, which is known for its famous historic George Town, a UNESCO World Heritage Site. It has attained UNESCO's Heritage City status and is exposed to artificial lighting consideration by Penang

A. A. S. Ali (✉) · S. A. Zakaria · A. C. K. Guan · C. J. Shun
Interior Architecture Department, School of Housing, Building and Planning, Universiti Sains Malaysia, 11800 Gelugor, Penang, Malaysia
e-mail: lolitta.usm@gmail.com

S. A. Zakaria
e-mail: ssafial@usm.my

M. Awang and M. R. Meor M Fared (eds.), *ICACE 2019*,
Lecture Notes in Civil Engineering 59, https://doi.org/10.1007/978-981-15-1193-6_13

State Government improvement plans [1]. Fortunately, people nowadays have started to notice these buildings and appreciate them. The government performed a lot of conservation works and renovation on heritage buildings instead of demolishing them. By doing this action, it has retained the uniqueness of the buildings; at the same time it has enhanced the aesthetic value of the heritage buildings. The Penang state government has also studied it, to measure the nature and extent of the building's lighting issue [1]. By adding lighting fixtures, the buildings become more noticeable during night time. Friebe (2007) stated that artists had opened people's eyes to a new way of comprehending things and relationships because lighting design has played a more or less subordinate role in architecture [2]. However, excessive lighting exposed during night time will lead to light pollution, which might bring harmful effects on human, animals, and plants. This paper demonstrates a case study of the lighting design applied on St. George's church in George Town in Penang Island. In this study, the researchers will further discuss some effects of lightings at night and also give some suggestions toward the lighting fixtures that can balance up the function and the aesthetics of the buildings.

2 Objectives

I. To analyze the lighting system of the heritage building.
II. To identify the issues caused by the lighting of the heritage building.
III. To propose a better lighting design for the heritage building.

3 Research Method

For this study, the researchers have decided to use a combination of three research tools, which are observation, measuring tools, and reading database. First, the observation method is used for collecting data about the lighting of the heritage building. In this method, the primary data was collected through site observation; the site is at St. George's Church in Georgetown, Penang. Researchers have visited the site at different timings during daytime and nighttime to observe and determine the differences in the condition of the building with daylight during daytime and artificial lighting during night time. Through this observation, the issue as well as the impact of the lighting was examined during different periods. Besides that, the characteristics of the visual environment as well as the types of light sources that applied on the exterior building or the surroundings of the building and the design arrangement of the lighting fixtures were reviewed. Second, the Lux meter was used to gauge illuminance of the lighting fixtures at different distance to the surfaces of the buildings in different periods. The data was collected and analyzed to explain the impact on the building and the effects on the environment. The quantitative method was used in collecting the data. Finally, the information was collected and the references including the research through library and Internet research were used to determine further explanation of this research. The sources to collect the information for this study vary, for example, journals, articles, books, papers, magazines and internet databases.

4 Description of the Heritage Building

The heritage building for this research is the St. George's Church, Penang (5°25′11.0″N 100°20′20.8″E) [3]. The building is the oldest Anglican Church in Southeast Asia, which is located in the historic city of George Town, Penang. The church is known for its Francis Light Memorial and one huge Mahogany tree located at the site. St. George's Church is one of the UNESCO's World Heritage Sites. It is located at the historical site of George Town with many heritage buildings around that area, such as Penang High Court Georgetown located opposite to it and Church of the Assumption, which is located along the same street.

The St. George's Church was built in 1816 and was completed in 1818 when Colonel J. A. Bannerman was the Governor of Penang. The St. George's Church was built with clay brick, stone, lime plaster, timber, marble, and terrazzo. Initially, St. George's Church was built with a flat roof, but then it was covered by a pitched roof using Marseille tiles due to roof leakage. This pitched roof has remained until today. Restoration works had been done on this building in 2010 after being diagnosed with numerous problems. It has been covered by a new coat of paint, which shows the elegance of the building.

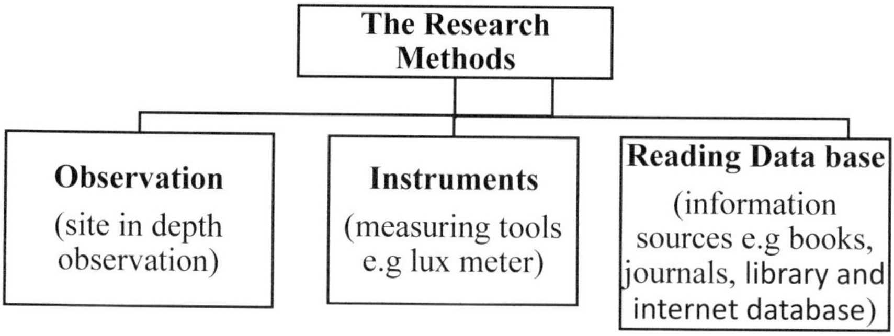

5 Findings and Discussion

5.1 Social Impact on the Implementation of the Lighting Fixture

The implementations of lighting fixtures onto historical buildings had given some impact to society. Lighting fixtures create a sense of drama and art to the building itself. Proper lighting fixtures help highlight the identities and history of historical buildings. In another way, this can benefit the country by growing the economy through tourism when the building becomes a landmark and attractive to tourists. However, an improper installation of lighting fixtures can be harmful to the society.

5.2 Primary Functions and Goals of the Lighting System

For most kinds of lighting systems, the luminous effect is determined by 4 factors: the lighting source's intensity, which depends on direction and characteristic; the geometry where the light interacts with the receiving surface; the receiving surface that modified the light to become the secondary light by reflection, which depends on redirection and coloring of light; and the person who views the lighting source and illuminated surfaces as he or she moves around [2]. The primary function of the lighting system is to strengthen the visibility of the structure of the building to the human eyes, especially during night time and low daylight conditions. Adding artificial lighting to the buildings helps attract the attention of the tourists. Besides that, lighting can affect the mood of people just by using different colors of light. Choosing proper colors of light can enhance the overall aesthetic value of the buildings. The lighting system also functions as guidance for people. For instance, the lighting system implemented in St. George Church along the corridor is to show a clear direction of the pathway to the people. Furthermore, lighting plays a vital role in enhancing the details of the heritage buildings through the shadow casting on the buildings which creates contrasts.

5.3 Types of Light Sources and Design Arrangements

The type of light source installed on St. George's Church is floodlight. The floodlights were located in front of every column. The placement of the floodlights was too close to each other, which caused the building to be overexposed. The high intensity of light reflected on the white surface of the building has caused glare to the human eyes. Floodlighting of buildings should be unobtrusive and set within the landscape and not attached to the building. Floodlights used on buildings must be natural white light.

5.4 Characteristic of the Visual Environment

Based on Fig. 1, the form and the shape of the heritage building of St. George's Church can be seen under the daylight, and the structures and textures of the building are highlighted during the daytime.

During the night time, the visual of the building is shown in Fig. 2. The lighting fixture has been arranged and appropriately installed in the form of a column by

Fig. 1 The building during day time under natural lighting

Fig. 2 The building during night time, under artificial lighting [4]

column. The installation of the lighting fixture does help to enhance a particular area of the building. However, the exposure of the lighting, which is projected on the white surface of the building has created unwanted glare that can give some impact to the environment.

5.5 Key Design Factors and Criteria of the Lighting System

The luminance level of the daylight is sufficient enough for the building to be visible. The uniformly distributed daylight is the most comfortable lighting to the human eyes. Natural daylight is hard to be compared the artificial lights. However, the presence of artificial light cannot be neglected as it is crucial when there is no daylight. Sufficient amount of artificial lighting can enhance the details of the building, but an excessive amount of light will cause light pollution.

According to the luminance level that has been collected using the lux meter, the luminance from the floodlight in front of St. George's Church is approximately around 30–40 lx. The researchers recorded the luminance level from different angles of the building and the luminance level was between the ranges mentioned above. From what the researchers observed, the floodlight was emitted from the ground toward the building; hence, high intensity of light is focused on the lower part of the building and gradually faded while moving upward.

For the type of light source, use low power input light source to reduce the intensity of the light. This is to avoid the cause of glare to our eyes and reduce light pollution to the environment. Light output could range from 2600 to 135,000 lumens (1 lx = 1 lm/m^2) [5]. Besides, a light source with high luminaire efficiency should be used in the building. The high efficiency of the luminaire can reduce the waste of energy consumed, while at the same time, it can reduce the cost. The finishing and appearance will also affect the lighting system. For St. George Church, the surface of the building is painted in pure white. The highly reflective white surface requires the comparatively low intensity of light to have a bright appearance. An excessive amount of light reflected from the white surface will cause glare. Lastly, for the color of the light source, warm color light should be used, which is less harmful to our health. This is due to the fact that a color with more cooling temperature consists of higher wavelength compared to a warmer color. White light is defined as the complete mixture of all of the wavelengths of the visible spectrum [6]. In fact, in the spectrum of light wavelength, if the color of a surface is anything other than white, it means that it absorbs light of some wavelengths [7].

6 Conclusion

In summary, lighting is an essential element in architecture. Buildings require both natural lighting and artificial lighting from the inside to the outside. This is due to the fact that the lighting is interconnected closely with our daily life. A good design of lighting can bring life and meaning to the building. It enhances the visual appearance, and the mood created by the buildings. The amount of light used depends on the intention of the designer. However, it can become an issue when the lighting is designed without consideration of health and social impacts. The extreme intensity of light will cause light pollution, which is harmful to the environment. Nowadays, technology has become much more advanced than in the past. There are artificial lightings, which are eco-friendly, which the designer should consider. However, a good architecture building design will never neglect the beauty of natural lighting. It is the most suitable and comfortable lighting to humans and the environment, which can never be completed by the artificial lighting.

References

1. Ariffin NFM et al (2015) Stakeholders' attitude on the willingness-to-pay value for the conservation of the George town, Penang world heritage site. J Surv Constr Prop 6(1):1–10. https://doi.org/10.22452/jscp
2. Deziel C (2017) Which colors reflect more light? Retrieved from https://sciencing.com/colors-reflect-light-8398645.html
3. Coriolano LMT, Tavares JM, Ateljevic I (2016) Solidarity economy as a counterpoint to classical economics: possibilities of changes. Int Multidisc J Tour
4. Irimia R, Gottschling M (2016) Taxonomic revision of Rochefortia Sw. (Ehretiaceae, Boraginales). Biodivers Data J 4: e7720. https://doi.org/10.3897/BDJ.4.e7720. (n.d.). https://doi.org/10.3897/bdj.4.e7720.figure2f
5. Zakaria SA et al (2018) Lighting heritage building practice in George town, Penang Island. In: IOP conference series: materials science and engineering, vol 401, p 012005. https://doi.org/10.1088/1757-899x/401/1/012005
6. Cardenas R (2015) White light: definition, source and spectrum. Retrieved from https://study.com/academy/lesson/white-light-definition-source-spectrum.html
7. van Santen C (2006) Light zone city—light planning in the urban context. Birkhaeuser, Basel, Boston, Berlin, p 10

Analysing Architecture: Dissecting the Malaysian "House"

Veronica Ng and Heffrence Teow

Abstract One of the areas of interest in architecture is the rhetoric of: What is contemporary Malaysian architecture? While there are publications that documented Malaysian architecture, the current state of documentation and research on recent and contemporary architecture suggests a need to take a more critical look into the design thinking and narratives of contemporary Malaysian architects and architecture. This paper contributes alternative views on recent and contemporary Malaysian architecture by examining the design thinking and narrative of Malaysian architects. Through interpretive research, this study analyzes eight houses completed between 1980s and the present, informed by a synthesis of architects' interviews, observations through fieldwork, and analytical and diagrammatic studies following the framework of Simon Unwin and Geoffrey Baker. The research argues that through dissecting architecture, the discursive relations between context, form, space and practice offers a critical approach to identify themes that define Malaysian architecture.

Keywords House · Malaysian architecture · Design thinking · Interpretive research · Diagramming

1 Introduction

Since the last few decades, the rise of postmodernity, which is premised on ideas of plurality and multiplicity, has informed a diverse palette of architectural outcomes. At present, the power of image via websites, such as Archdaily, Contemporarist, and Dezeen, and web apps, such as Pinterest provides unlimited access to visually powerful images of architecture, and hence, forming the backdrop to contemporary architecture, which is succumbed to the influences of global language, a language for "all" rather than one that is specific to its context. The influence of the global image is further compounded with the lack of publications on context-specific architecture, particularly in Malaysia, and the architecture education that teaches us Western history. As such, architecture is being reduced to images, rather than the intellectual,

V. Ng (✉) · H. Teow
School of Architecture, Building and Design, Taylor's University, Subang Jaya, Malaysia
e-mail: Ng.foongpeng@taylors.edu.my

M. Awang and M. R. Meor M Fared (eds.), *ICACE 2019*,
Lecture Notes in Civil Engineering 59, https://doi.org/10.1007/978-981-15-1193-6_14

the contextual, and the narration of meaning that is an outcome of the explorative process of design. Despite the diversity and uniqueness, there are limited publications and studies that question the intellectual contexts and complexity of Malaysian architecture.

2 Literature Review

Within the Malaysian architecture scene, one of the key moments of clarity on the definition of Malaysian architecture was the architecture of nation building during the 1960s to the 1980s by the first generation of Malaysian architects and the pioneers of local Malaysian architecture, witnessed through buildings, such as the National Stadium, the Parliament House and among others. Recent publications on Malaysian architecture, particularly in its formative years and on modern architectural heritage, named as Shapers of Modern Malaysia [1], The Living Machines: Malaysia's Modern Architectural Heritage [2], Architectural Heritage (Kuala Lumpur Pre-Merdeka) [3], Building Merdeka: Independence Architecture in Kuala Lumpur 1957–1966 [4]. To date, publications on recent and contemporary Malaysian architecture reflect several directions. First, publications on Malaysian houses and how they are cultural and tropical responsive, such as Recent Malaysian Architecture [5], Tropical Style: Contemporary Dream Houses in Malaysia [6], collection of contemporary houses in the New Malaysian House [7], Rethink—A New Paradigm for Malaysian Timber [8], and design showcase of luxury Asian homes in Sustainable Asian Houses [9]. Second, publications on architects' portfolio of works through monographs, such as Homes, Issues + Processes: design collective architects [10], ATSA Monographs [11], and anthology such as Malaysian Emerging Architects: 30 Under 40 [12]. Third, documentation of Malaysian architecture in the form of exhibition/exhibition catalogs, published books, journals and magazines, namely, Architecture Malaysia, the official magazine of the Malaysian Institute of Architects. While the publications document the breath of Malaysian architecture, there are opportunities to develop an understanding of recent and contemporary Malaysian architecture. Through these publications, the underpinning architectural thinking is viewed as a backdrop, rather than playing a pivotal role in the production of architecture.

Anchoring from the research standpoint, the question of identity is central, particularly the rhetoric of "what is Malaysian identity?" In the book Malaysian Architecture: Crisis Within, Professor Dr. Mohamad Tajuddin commented on the lack of research about Malaysian architecture. Tajuddin [13] critically questioned: Do we have an architectural identity? What future does a Malaysian architectural identity have? Drawing from history to the postmodern, he suggested the presence of Malaysian identity and he classified the diverse approaches toward identity-making in Malaysian architecture, alluding to the nature of identity as one, which is diverse rather than Malay-centric, and the notion that Malaysian identity is portrayed in myriad ways, such as through revivalist approach, primitive regionalism, ethnic symbolism, and tropical heritage [13] (p. 18). Coming from similar basis, the recent book titled Theorizing Emergent Malaysian Architecture [14], theorized works of 10 architectural practices, and argues for diversity of approaches in defining emerging approaches towards Malaysian architecture.

The current state of documentation and research on recent and contemporary architecture suggests a need to take a more critical look into the design thinking and narratives of contemporary Malaysian architects and architecture. Moving away from the documentation approach, and the periodization as well as the identity discourse, this research calls for the need to critically examine architecture, in order to reveal the principles that drive the design thinking and narrative of the architect. It aims to identify alternative views on recent and contemporary Malaysian architecture through theorizing projects. Through analyzing architecture, as Unwin [15, 16] argued, one is enabled to make sense of the complex and diverse aspects of architecture.

Within this discourse, the typology of "house" appears to be a relevant object for study. The design thinking of architects are often experimented and expressed through the design of the "house", either the architect's house or a house, which is representative of the underpinning philosophy and position of the architect; For example, Le Corbusier and the Villa Savoye, Mies van der Rohe and the Farnsworth House and Robie House, Robert Venturi and the Venturi House, and Peter Eisenmen and House I-X, among others. The house forms an ideal object for the manifesto of architectural ideals due to its complexity albeit its relatively intimate scale compared to other building types, which allows for critical and thorough analysis.

In its pragmatic sense, the "house" is indeed one of the most common building typologies in architecture. Theoretically, the notion of the "house" is a rhetorical question, which appeared in historical and contemporary discourse on architecture. What is a "house"? Why the problem of the house? Amidst the transient nature of contemporary life, the explorations of the house have the propensity to offer ways of defining architecture of today. The discourse on "house" is interesting because the house is a theoretical concept, as well as an object of architecture, which is most "patronized". Taking this as a point for departure, this research examines the design thinking and narrative of Malaysian architects by using the "house" as object for analysis. In order to achieve the aim, this study analyzes the house as an architectural construct, and theorizes it as an architectural form, which reflects the design thinking and narrative of the architect.

3 Materials and Methods

Through interpretive research, this study proposes to analyze eight (8) houses through a cross section of time:

1. Salinger House (1988)—Jimmy Lim;
2. Roof-Roof House (1991)—Kenneth Yeang, T.R. Hamzah and Yeang;
3. Wooi Residence (2004)—Wooi Lok Kuang, Wooi Architect;

4. M + M House (2009)—Mike Boon, Arkitek JFN;
5. S11 (2013)—Dr. Tan Loke Mun, Archicentre;
6. 38Mews (2015)—Chan Mun Inn, design collective architects;
7. Chempenai House (2015)—WH and BC Ang, WHBC Architects;
8. Twinkle Villa (2016)—Chan CY, CY Chan Architect

These houses are combinations of architects' houses as well as representatives of the philosophy of the selected architecture practice. To limit the scope of research, the criteria for the selection include: The houses are designed by Malaysian registered architects; The houses have gained international/national recognitions through awards or publications; The architects are known regionally or globally; The houses must be completed within the timeline of 1980s to the present.

Qualitative by nature, the interpretation is informed by a synthesis of interviews carried out with the selected architects, and observations through site visits and analytical and diagrammatic studies on the selected houses. Interviews and observations through site visit offer the experiential aspect of the houses. The interviews are conducted to collect data on the philosophies and the background of the selected architects, and the process of design, while the site visit will offer data on the experiential qualities of the space and the relationship of the architecture to its contextual and cultural conditions of the site. Subsequent to the interview and observations, the eight houses will be analyzed by using the frameworks adapted from Simon Unwin [15, 16] and Geoffrey Baker [17]. In order to "make sense of the complex and diverse aspects of architecture", Unwin [15, 16] analyzed architectural ideas developed and shown through drawings, where interrogating the published materials is part of the analytical process; And redrawing the plans and sections of buildings under scrutiny is an essential part of that interrogation. Starting from Unwin's first analytical book, *Analyzing Architecture* to the *Twenty-Five Buildings Every Architecture Should Understand*, the conceptual framework of analysis are as follows: identification of place; basic elements; modifying elements; primitive place types; temples and cottages; geometries of being; ideal geometry; stratification; space and structure; parallel walls, transition, hierarchy, heart; and others.

In this research, these aspects of analysis are conducted via diagramming which draws precedence from Baker's *Le Corbusier, An Analysis of Form*. Baker's study focused on architectural organization by examining Le Corbusier's manipulation of form. The analysis on the aspects of form examined the site forces, centroidal and linear form; dynamics of form (core systems, linear systems, axial systems, enchelon and radial systems, and interlocking systems), and form distortion. In order to do this, a diagrammatic method of analysis is used, which dissects the form in order to show how the various elements are related to each other and to particular site conditions, and to reveal certain principles of design that can be discerned in Le Corbusier's works.

Both Unwin and Baker's works form a theoretical framework to study the eight houses, with the use of diagrammatic methods of analysis. This analytical approach, in addition to the interview and site visit, offers formal and experiential means of understanding about where the architect's design thinking (where do ideas) come from, and how are they executed into tangible built forms and environments. Subsequently, the analyses of the eight houses will be compared to identify themes. These themes will be discussed and relationships will be drawn in order to identify an alternative way of understanding the Malaysian architecture.

4 Discussions

The implications of the findings point toward the house as an architectural construct emerging from the discursive formation of idea. The house is more than a material form; it is a discourse in itself. The findings echoed the concluding words by Unwin [16], who argued that the building is a medium. "Physically, sensually, psychologically, socially and emotionally, and architecture is without doubt, the richest of all the arts." [15] (p. 269)

Based on the data collected through field works, interviews, and literature, each house has its central characteristic that defines its architectural construct.

- Salinger House redefines the ideas of vernacular architecture/tropical architecture
- Roof-roof House is a modern interpretation of tropical house based on bio-climatic architecture
- Wooi's Residence is an interpretation of individual and collective spaces based on the notion of enclosure and intimacy
- M + M House is about the idea of the collective and the open.
- S11 is an interpretation of tropical resort living, and an exploration of Green Building Index (GBI) idea
- 38 Mews is about the idea of an oblique modern form.
- Chempenai House is about the negotiation between construction/structure, site, and client.
- Twinkle Villa is about a dialog between form, site, and space. The site becomes a powerful tool that choreographs the form.

While the house has been an object for analysis in this study in the above analysis, the architect's design thinking and narrative is central to the production of the house as an artefact—a discursive object within the discourse on architecture. This is aligned to and extends discussion of the house in current book publications, positioning the house not only as an artefact, which is contextual to Malaysia, but also as an architectural typology that is positioned as an idea and practice.

The implication of the findings derived from the eight houses suggests that the house is an architectural typology that has diverse interpretations; the commonality is that the house becomes impetus to design thinking that drives the architectural practice of these architects. Although the houses are from the period of 1980s to the present, the findings suggest the importance of the house as a common object of architectural discourse in relation to practice.

- Yeang sees the house as a "machine" toward an exploration of bio-climatic system.
- Jimmy Lim sees the house as a typology for exploration of tropical and vernacular architecture, the common sense of responsive architecture within its context.
- Wooi sees the house as a sacred typology.
- Boon views the house as a simple form restraint from excessive embellishments, very much simplified to the notion of shelter (metaphoric of a chicken coop).
- Tan Loke Mun views the house as an experimentation of idea and ideals of green architecture, and extension of exploring resort living.
- Chan reflects that the house is about one's personal ideal and it relates to customization and personalization.
- Ang B.C. views the house as a "house", without much association to theoretical underpinning, but rather the task is to find inconsistency in the contexts, rather than working with a prescribed set of "habits", which may lead to egoistic architecture. The house is indeed a modest answer to the question related to the "problem" set for each house.
- While C.Y. Chan's practice focuses more on public buildings and high-rise architecture, the house is an emerging form that enables expressions of the individual client.

However, the relationship between the house and the design thinking is not direct. The synthesis of the findings illustrates that both design ideas and design thinking/narratives of architects are discursive by nature; i.e., they are inter-related between different conditions that shape architectural space, structure, and form. Within this web of relations, the house is a discursive object positioned within the broader discourse on contemporary Malaysian architecture.

5 Conclusions

The findings of this study illustrate the richness and many dimensions of architecture—there is more to architecture than physical appearance. These analyses illustrate that there is no single way to do architecture. But these analyses also show that buildings designed according to different attitudes, techniques, and methodologies can be studied using a consistent conceptual framework (when used selectively rather than a checklist) to define the underlying architecture of particular buildings.

At the outset, the houses chosen for analysis are those of architects or closely reflect the architect's body of works throughout 1980s to the present. From this, consistent yet diverse themes toward the design thinking and narrative of Malaysian architects emerged: typological, psychological, space, personal, explorative, and methodological. These themes emerge from a network of discursive relations between the built form, design thinking, and narrative of the architect, and other generative forces of architecture. Synthesizing the themes, there appears to be three aspects to the design thinking relating to space, person, and methodology (practice).

This interpretation expands the understanding on contemporary Malaysian architecture. Firstly, it reinforces the notion of architecture beyond a physical object; more so, it should be understood beyond the descriptive stance. As much as the study of architecture, this study argues for a critical look at the nexus among architecture as an object, idea, and a practice—repositioning of the object/building/architecture as part of the discursive formation of the discourse on contemporary Malaysian architecture rather than merely a product that represents it.

While the analysis reveals themes of design thinking and analysis of the house, it suggests that in order to understand contemporary Malaysian architecture, the house (object) should be viewed as part of the discourse of contemporary Malaysian architecture. From the critical study of the design thinking and narratives of contemporary Malaysian architects through the eight houses, the findings showed the relationship of generative and discursive forces in relation to the production of architecture.

It casts light on alternative views on recent and contemporary Malaysian architecture, viewing architecture as a discursive artefact that defines and is defined by the intangible conditions, particularly the architect as an agent of the production of architecture. It means, when we study architecture, we should study the relationships rather than the outcome. Future research may pursue on the analysis of architectural object as part of the discourse on contemporary architecture.

Acknowledgements This research is funded by Taylor's Research Grant Scheme from 2017 to 2018

References

1. Ngiom L (ed) (2010) Shapers of modern Malaysia. Pertubuhan Akitek Malaysia, Malaysia
2. Ariffin K, Hussain, N, Ng FP (2015) The living machines: Malaysia's modern architectural heritage. In: Liew VF (ed). Pertubuhan Akitek Malaysia, Malaysia
3. Pertubuhan Akitek Malaysia (2010) Architectural heritage (Kuala Lumpur Pre-Merdeka) Pertubuhan Akitek Malaysia, Malaysia
4. Lai CK (2007) Building Merdeka: independence architecture in Kuala Lumpur 1957–1966. Petronas, Malaysia
5. Goad P, Ngiom L (2007) Recent Malaysian architecture. Pesaro Publishing, Australia
6. Beal G, Termansen J (2008) Tropical style: contemporary dream houses in Malaysia. Periplus Editions, Hong Kong
7. Powell R (2008) The new Malaysian house. Periplus Editions, Hong Kong
8. Lee CW, Zainal S (2012) Rethink—a new paradigm for Malaysian timber. MPH Publishing/Malaysian Timber Council, Malaysia
9. McGillick P, Kawana M (2017) Sustainable Asian houses. Tuttle Publishing, North Clarendon, VT
10. Cheah F (ed) (2011) Homes, issues + processes: design collective architects. Black Dog Publishing, London
11. ATSA monograph. http://www.atsa.com.my/folio_monograph.html. Assessed 20 July 2016
12. Emerging Malaysian architects: 30 under 40. Pertubuhan Akitek Malaysia, Malaysia
13. Tajuddin M (2005) Malaysian architecture: crisis within. Utusan Publications, Malaysia
14. Ng V (2016) Theorizing emergent Malaysian architecture. Pertubuhan Akitek Malaysia, Malaysia

15. Unwin S (2014) Analyzing architecture, 4th edn. Routledge, New York and London
16. Unwin S (2015) Twenty-five buildings every architect should understand. Routledge, New York and London
17. Baker G (1996) Le Corbusier, an analysis of form, 3rd edn. Van Nostrand Reinhold, New York

Assessment of Building Façade Thermal Performance for the Potential Application of Phase Change Materials (PCMs) in Malaysia, a Case Study

Zeyad Amin Al-Absi, Mohd Hafizal Mohd Isa and Mazran Ismail

Abstract Phase change materials (PCMs) can be applied in buildings to reduce daily temperature fluctuation and shift the peak temperature occurrence, hence, providing more thermally comfortable environments. This approach helps in reducing the energy consumption associated with the use of mechanical systems. The main concept of PCMs is to absorb the extra ambient heat during the day time and reuse or remove it during the night time. Therefore, it is crucial to have an ambient temperature fluctuation in order for PCMs to efficiently work. This paper investigates the potential application of PCMs in building façade within tropical climates to reduce the overheating due to the direct solar radiations received by building façade. A building was selected to conduct measurements on its southern and northern opaque façades during the period in which the sun is facing the southern façade. Measurements included the external surface temperature (T_{se}), internal surface temperature (T_{si}) and heat flux of the selected facades, as well as the indoor air temperature (T_{ai}) and global solar radiation (GSR). It was found that the southern façade was highly influenced by the GSR and reached maximum T_{se} of 44.81 °C compared to 32.63 °C for the northern façade, which was almost constant through the measurement period regardless of GSR. In addition, the effects of the high T_{se} on increasing T_{si} and, then, T_{ai} was clear. Furthermore, high daily fluctuation of T_{se} was observed that reached up to 18 °C, which might propose a potential area for the application of PCMs to reduce the overheating caused by the direct solar radiation. Finally, analysing the T_{se} of the southern facade during the sunniest days indicates that the optimum transition temperatures for the PCMs selected for the application in the investigated conditions could be between the range of 30–36 °C.

Z. A. Al-Absi (✉) · M. H. Mohd Isa · M. Ismail
School of Housing, Building and Planning, Universiti Sains Malaysia, 11800 George Town, Pulau Pinang, Malaysia
e-mail: zeyadarch@gmail.com

M. H. Mohd Isa
e-mail: hafizal@usm.my

M. Ismail
e-mail: mazran@usm.my

M. Awang and M. R. Meor M Fared (eds.), *ICACE 2019*,
Lecture Notes in Civil Engineering 59, https://doi.org/10.1007/978-981-15-1193-6_15

Keywords Phase change materials (PCMs) · Thermal energy storage (TES) · Building façade · Direct solar radiation · Surface temperature

1 Introduction

Nowadays, buildings' designs have become more than aesthetics and space requirements. They must be designed in such a way that they react with the surrounding environment [1]. For instance, building fabric should be selected and designed to provide more thermal stability for the internal environment in order to prevent the high usage of mechanical cooling/heating systems, which is associated with high energy consumption, causing buildings to account for 40% of global energy consumption [2].

Passive design strategies utilize the energy available from the natural environment to be used in buildings for heating and/or cooling purposes. This is achieved by heat absorption, prevention or dissipation based on the surrounding environment [3]. The ability of the building's fabric to improve air temperature and thermal comfort of the internal spaces by exchanging the heat is known as the thermal performance of a building [1]. Passive design strategies are promising alternatives to conventional heating and cooling systems [3] and may become more helpful for some types of buildings (i.e. residential buildings and some small commercial buildings), which are associated with the low density of lighting, equipment and occupants [4]. Since building envelope plays a key role in controlling the effects of external weather through the control of heat transmission into the building, it is linked with many passive techniques that address wall and roof insulation, fenestration areas and the type of glazing [4].

Thermal energy storage (TES) is a useful sustainable passive technology that can be used in building fabric to improve heat exchange and energy efficiency and to minimize energy consumption [5]. TES can store the available unwanted heat and reuse it once needed, which handles the mismatch between the energy's supply and demand [2]. Heat can be stored in three types: sensible heat, latent heat and thermochemical of which, the latent heat is the most efficient that stores a higher density of thermal energy with smaller changes in temperature. Materials that can store the latent heat through a change in their phase are known as phase change materials (PCMs) [3, 6]. PCMs have proven their effectiveness in lowering the indoor temperature, providing thermally comfortable environments and reducing energy consumptions [2].

Different types of PCMs that have a wide range of transition temperatures are available. Selecting PCMs' transition temperature range is influenced by climate, system type for PCMs application or its position in the building [7]. In addition, the optimal transition temperature for PCM layers is influenced by the temperature of the ambient environment [8]. PCMs work by changing their phase from solid to liquid and then back to solid. Achieving a daily complete thermal transition cycle is crucial for the efficient performance of the PCMs. Hence the ambient temperature is

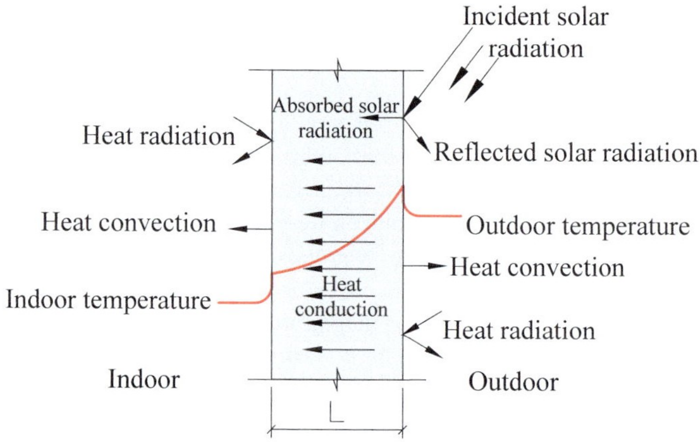

Fig. 1 Heat transfer in building opaque façade adopted from [10]

required to have a temperature fluctuation in which the PCMs can efficiently work and reduce this temperature fluctuation. According to Khadiran et al. [7], the diurnal temperature fluctuation must be suitable to allow a full thermal cycle of the PCMs. In addition, it was suggested to be more than 10 °C in order to ensure complete PCM transition cycles and effective performance [9]. The indoor environment of different buildings with different orientations have been evaluated as part of this research. However, the results showed a limited indoor air temperature fluctuation with an average of 2 °C to 4 °C, which indicates unsuitability of the indoor environment for the application of PCMs.

On the other hand, building façade receives the direct solar radiations and is considered the first protection aid for the indoor environment against the outdoor extreme weather conditions. Building façade receives the heat on the external surface by radiation and convection, transfers it by conduction through its section to the internal surface, then, by radiation and convection to the indoor environment (Fig. 1). Therefore, in this paper, the effects of the direct solar radiations received by building façade on raising its external and internal surface temperatures and indoor air temperatures due to the heat transfer through building façade were investigated. This investigation looks for the potential of PCMs' application in the external surfaces of building façade to reduce the overheating due to direct solar radiation.

2 Methodology

This work required conducting field measurements on building façades with different orientations to compare two scenarios: façade exposed to the direct solar radiation and façade not exposed to the direct solar radiation. Based on the sun path of Penang, the sun reaches the farthest point to the south on 21 December and the southern façade receives the maximum direct solar radiations while the northern façade is in the shade. Therefore, the measurement period was planned to start at this point and onward. For the measurements purpose, an apartment with three bedrooms at N-Park condominium, Penang, Malaysia, was selected, in which one of the bedrooms has an external wall orientated towards the south, while another bedroom has an external wall oriented towards the north. Therefore, these two bedrooms were chosen to conduct the required field measurements on the southern and northern facades. The wall was constructed from 100 mm of concrete and 10 mm of cement plastering.

The required parameters in this work are external surface temperature (T_{se}), internal surface temperature (T_{si}), heat flux, indoor air temperature (T_{ai}) and global solar radiation (GSR). The instrument used in this work was BABUC/M data logger with the required sensors (i.e. surface temperature, heat flux, air temperature and global solar radiation sensors) manufactured by LSI LASTEM, Italy. Two surface temperature and one heat flux sensors were used and attached to the centre of the opaque part of the façade (i.e. one surface temperature sensor on the external surface of the wall and the other one, with the heat flux sensor, on the internal surface of the wall, opposite to the external sensor). The T_{ai} sensor was placed in the middle of the room at 1 m above the ground based on the ASHRAE standard requirements [11]. Finally, the GSR sensor was placed 1 m away from the southern façade to the outdoor environment. A scheme for the measurements' setup is presented in Fig. 2.

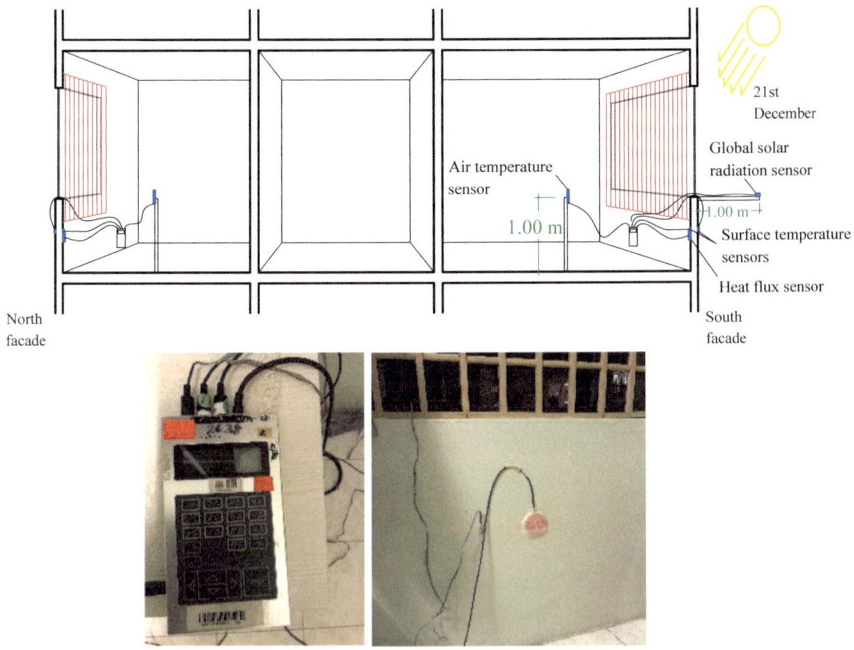

Fig. 2 Scheme for the measurements' setup (up), instrument used (down left) and sensors installation (down right)

Due to the limited inputs of the data logger, not all parameters can be measured at the same time. Therefore, the measurements of T_{se}, T_{si}, heat flux and T_{ai} were conducted first for the southern façade then for the northern façade. After that, measurements of T_{se} and GSR were conducted for both facades at the same time for comparison. Furthermore, because of the unstable weather conditions and the continued presence of cloudy weather periods, the measurement in each case lasted for a period that ensures getting at least 6 days with mostly sunny conditions (i.e. continuously or in two separated periods). Finally, curtains were left closed all the time during the measurement periods to reduce the effect of direct solar radiation penetrating through the windows and, in addition, the two bedrooms were unoccupied and closed during the measurement period to limit the effect of internal heat gain.

3 Results and Discussion

Field measurements in this work were conducted within the period of 21 December 2018 until 5 February 2019. Since the objective of this work is to investigate the effects of the direct solar radiation on overheating of building through its façade, measurements presented here were selected from the periods in which the sky condition was mostly clear and sunny.

3.1 Southern Façade

Measurements for the southern façade were conducted from 21 December 2018 to 31 December 2018. Figure 3 presents T_{se}, T_{si} and T_{ai} while Fig. 4 presents the heat flux. Figure 3 clearly shows the sharp increase in T_{se} and T_{si}. The T_{se} reached the highest peaks ranged 41.84–44.81 °C during the sunniest days while the lowest peaks ranged 35.14–35.98 °C during the cloudiest days. In addition, the everyday fluctuations on the peaks of T_{se} can be noticed, which can be linked to the continued change of the sky condition that affects the intensity of direct solar radiations on building façade. The maximum, minimum and average recorded T_{se} were 44.81 °C, 25.65 °C and 31.83 °C, respectively. However, the average recorded T_{se} increased up to 32.85 °C if selecting the sunniest days only. Moreover, the daily fluctuation of T_{se} was on average 10 °C during the cloudy days and increased up to 18 °C during the sunny days.

Additionally, the effects of T_{se} on T_{si} can be clearly shown, which are linked to the heat being conducted through the wall section. T_{si} is following exactly the same trends of T_{se}, and reached the highest peaks ranged between 37.73 and 39.55 °C during the sunniest days compared to 32.63–33.35 °C during the cloudiest days. The maximum, minimum and average recorded T_{si} were 39.63 °C, 26.29 °C and 31.40 °C, respectively. On the other hand, T_{ai} has lower peaks compared to both T_{se} and T_{si}, nevertheless, it has the same trend. For instance, high T_{se} and T_{si} resulted in higher

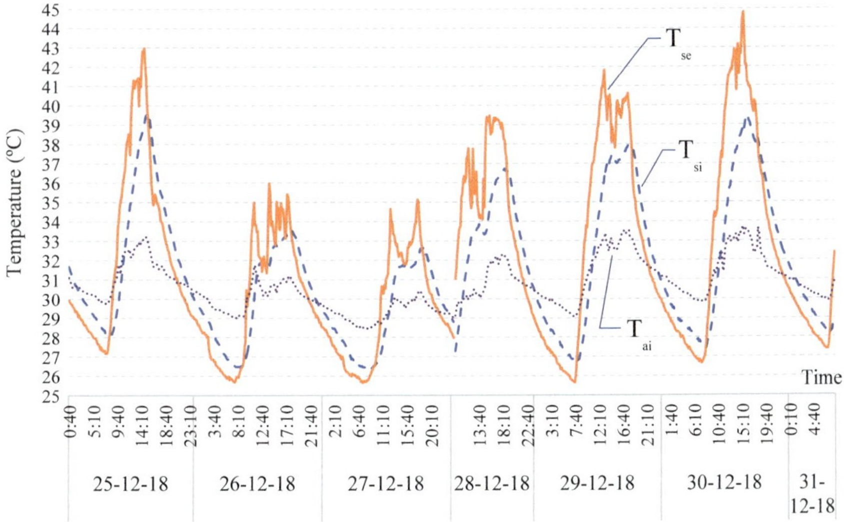

Fig. 3 T_{se} and $T_{si,}$ for the southern façade and T_{ai} (i.e. façade was exposed to direct solar radiation)

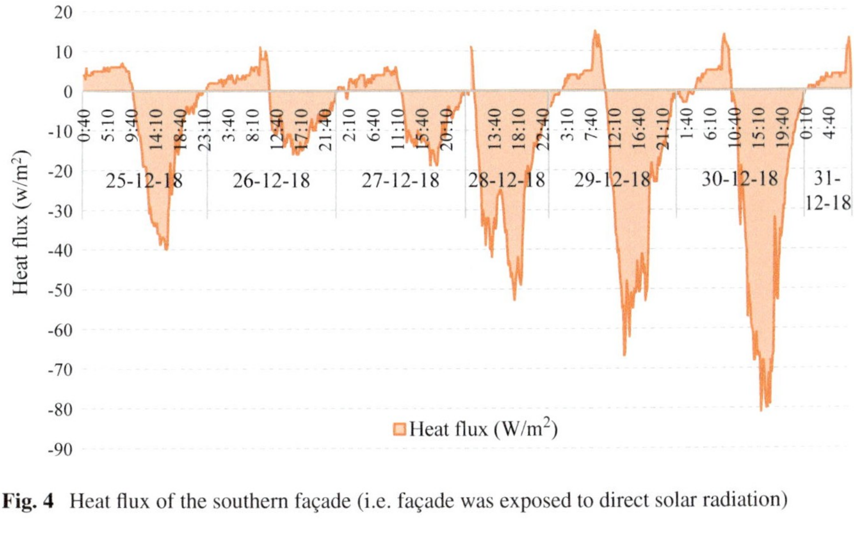

Fig. 4 Heat flux of the southern façade (i.e. façade was exposed to direct solar radiation)

T_{ai}. It is true that heat can still penetrate through the window even if the curtains were closed. However, it also may be attributed to the heat being transferred from the hot internal surface of the wall to the indoor environment by convection and radiation as mentioned earlier. The highest peaks of T_{ai} ranged between 32.21 and 33.73 °C for the sunny days compared to 30.28–31.76 °C for the cloudy days. In addition, the maximum, minimum and average recorded T_{ai} were 33.73 °C, 28.42 °C and 30.69 °C, respectively. In contrast, the minimum records for T_{se} and T_{si} were close and the

T_{si} was always higher by an average of 1 °C compared to T_{se}. However, minimum T_{ai} was always higher by 3–4 °C compared to T_{se}. This might be attributed to the absence of effective ventilation and to the absorbed heat in the wall being released during the night time that might increase the indoor air temperature.

The heat flux in Fig. 4 also demonstrates the effects of direct solar radiations on building façade in overheating the indoor environment. The negative value of the heat flux indicates that the direction of the heat flow is towards the indoor environment while the positive value is towards the outdoor environment. It can be seen clearly that during the sunny days, the heat flux reached up to 81 w/m^2 compared to 16–18 w/m^2 for the cloudy days. This figure clearly illustrates the high amount of heat transferred through building façade to the indoor environment during the day time and, in contrast, the low amount of heat that transferred back to the outdoor environment during the night time (i.e. 6–15 w/m^2). This heat, if not being treated well, might contribute to the overheating sensation and cause thermally discomfort conditions. It will also increase the cooling loads if air-conditioning systems are being used resulting in higher energy consumption.

3.2 Northern Façade

Measurements for the northern façade were conducted on 1, 2, 3, 7, 8 and 9 January 2019. Figure 5 presents T_{se}, T_{si} and T_{ai}, while Fig. 6 presents the heat flux. The limits

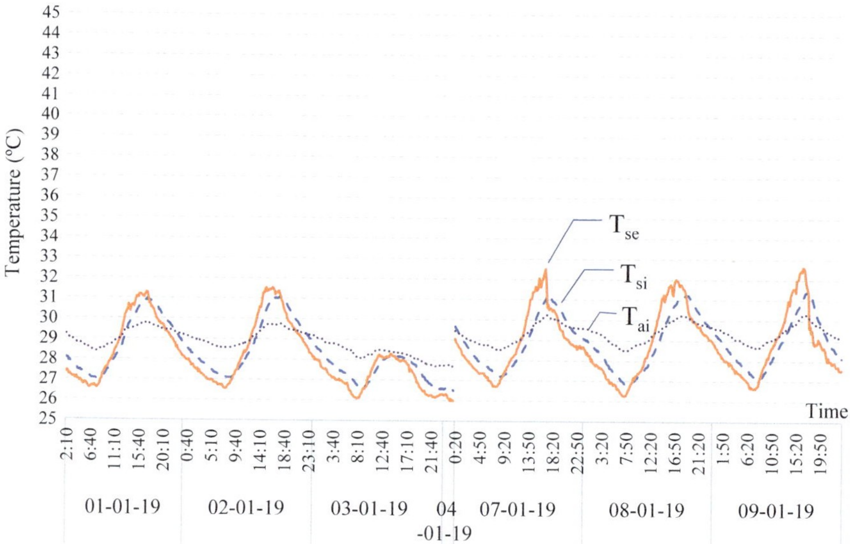

Fig. 5 T_{se} and T_{si}, for the northern façade and T_{ai} (i.e. façade was not exposed to direct solar radiation)

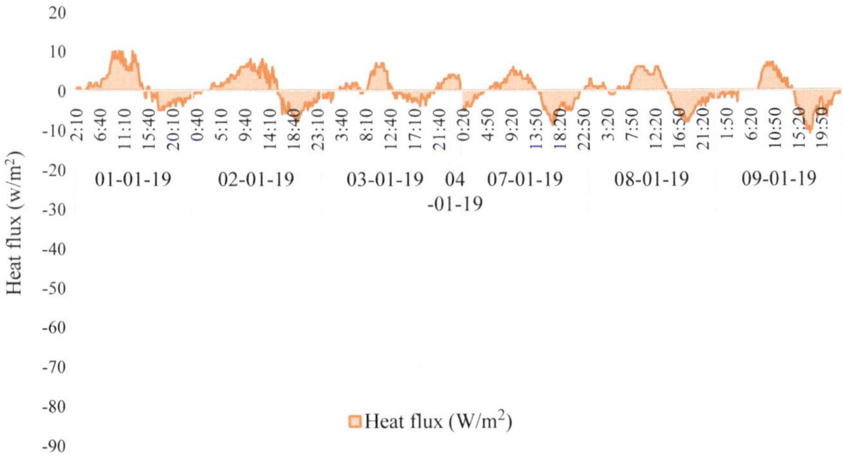

Fig. 6 Heat flux of the northern façade (i.e. façade was not exposed to direct solar radiation)

of the vertical axis in these figures (i.e. temperature and heat flux) were set similar to the limits in the southern façade's figures for comparison purposes. It can be seen clearly, in Fig. 5, that the highest peaks of T_{se} ranged between 31.7 and 32.52 °C regardless of the weather conditions (i.e. sunny or cloudy) except for the third day, in which it drops to 28.42 °C, which might be attributed to the rainy weather. The maximum, minimum and average recorded T_{se} were 32.63 °C, 25.99 °C and 28.48 °C, respectively. Furthermore, the daily fluctuation of T_{se} ranged between 5 and 6 °C for most of the days.

Furthermore, T_{si} also showed the same trends as T_{se} with a difference not more than 1 °C less than the maximum T_{se} or higher than the minimum T_{se}, thus, its daily fluctuation was ranged between 4 and 5 °C. Additionally, the maximum, minimum and average recorded T_{si} were 31.46 °C, 26.48 °C and 28.63 °C, respectively. On the other hand, T_{ai} had a very limited daily fluctuation that is less than 2 °C while its maximum, minimum and average records were 30.32 °C, 27.62 °C and 29.10 °C, respectively.

Figure 6 reflects how low heat is transferred to the indoor environment through the northern façade that is in the shade and does not receive direct solar radiations. The maximum recorded heat flux towards the indoor environment was 11 w/m^2 while it was 10 w/m^2 towards the outdoor environment. It can be noticed that almost the same amount of heat transferred to the indoor environment through building façade during the day time transferred back to the outdoor environment during the night time.

3.3 *Comparison of Southern and Northern Façades with Global Solar Radiation*

Measurement of T_{se} and GSR for both southern and northern façades was conducted within the period 26 January–5 February 2019 and is presented in Fig. 7. It can be seen clearly that T_{se} of the southern façade follows the same trends of the GSR, while, in contrast, T_{se} for the northern façade is mostly constant and does not follow the GSR. For instance, the maximum daily GSR recorded in most days ranged between 902 w/m^2 and 1007 w/m^2 except for 1 February, in which it dropped to 536 w/m^2, which can be attributed to the weather condition on that day. Similar trends were found with T_{se} of the southern façade, in which a high GSR resulted in high T_{se} of the façade and vice versa. The maximum daily T_{se} recorded from the southern façade ranged between 36.61 and 39.16 °C while the lowest record was 33.83 °C on the same day that received the lowest record for maximum daily GSR. In contrast, the maximum daily T_{se} recorded for the northern façade ranged between 30.03 and 31.63 °C for the whole period of measurements regardless of the GSR trend. It can be said that the T_{se} of the northern façade is influenced only by the outdoor air temperature in the current case (i.e. high rise building without any nearby buildings that might reflect solar radiation to the northern façade).

Furthermore, the minimum daily T_{se} recorded for both facades was almost constant and the difference between the highest and lowest records was 1.10 and 1.33 °C for the southern and northern facades, respectively. In addition, the minimum daily

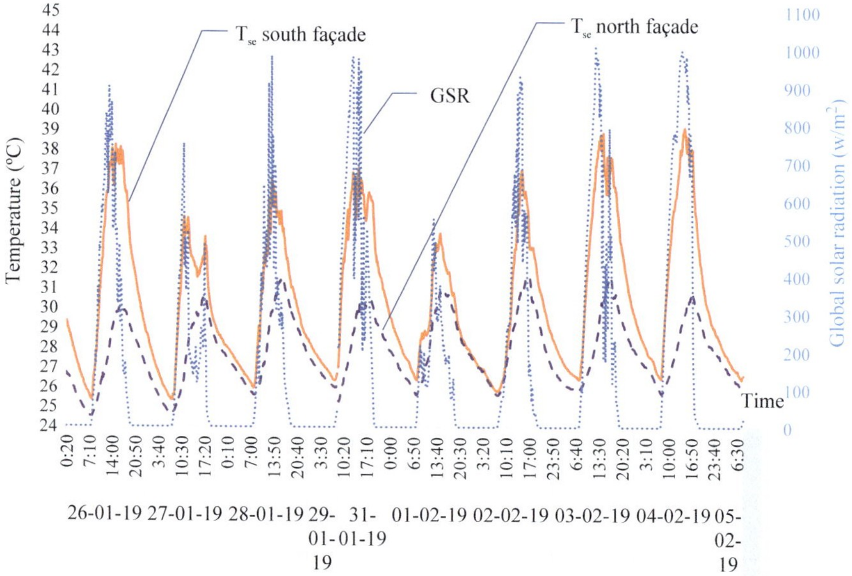

Fig. 7 T_{se} for the southern and northern facades with comparison to the GSR (i.e. southern façade was exposed to direct solar radiation while the northern façade was in the shade)

T_{se} records of the southern façade were higher than the northern façade, which might be attributed to the higher T_{se} and T_{si} of the southern façade during the day time that take longer time during the night time to release the heat to the ambient environment. It is worthy to mention here that the T_{se} of the southern façade gradually decreased since the start of the measurements in December to the end of measurements in February and beginning of March. This is linked to the sun path that reaches the farthest point to the south in the sky on 21 of December and then starts going back to the north.

3.4 *Potential of PCMs Application on the External Surface of Building Facade*

Phase change materials (PCMs) work by reducing temperature fluctuation and the peak temperature, and delay the time of peak temperature, which contributes to a better thermal condition. It was suggested that the ambient temperature fluctuation must be more than 10 °C in order to ensure a complete PCMs transition cycle and effective performance [9]. Based on what was discussed earlier, building façade that was exposed to direct solar radiation reached an extreme T_{se} close to 45 °C, which resulted in higher T_{si} and T_{ai} compared to building façade that was not exposed to direct solar radiation. In addition, the daily fluctuation of T_{se} reached up to 18 °C compared to only 6 °C for the exposed and non-exposed façades to direct solar radiation, respectively.

The application of PCMs in building materials increases their thermal mass. In the current case, the need for storing the heat is to prevent its penetration to the indoor environment during the day time and release it back to the outdoor environment during the night time. Theoretically, applying PCMs in the external layer of building façade will increase its thermal mass and help to absorb the heat coming from the received direct solar radiation during the day time within the PCMs. This results in less heat being conducted to the internal surface of the façade (i.e. less T_{si}). Therefore, less heat will be transferred to the indoor environment, which will reduce the T_{ai} and produce better indoor thermal environment. The amount of PCMs used in the external layer of the façade will define how long it can absorb the heat before it completely melts resulting in the heat conducting to the indoor environment. On the other hand, the transition range of PCMs (i.e. melting and freezing temperatures) must be selected carefully to achieve the best result and ensure daily complete melting/freezing cycle.

In Fig. 8, the three sunniest days from the measurement period that were conducted at the end of December and the three sunniest days from the measurement period that were conducted at the end of January were combined in one graph in order to find out the PCMs transition range that can ensure complete daily melting/freezing cycle. Based on [12], selected PCMs in many previous works had transition temperatures close to the average ambient temperature in which the PCMs melt and freeze. The average T_{se} for the selected days is close to 32.5 °C. In addition, to select the transition

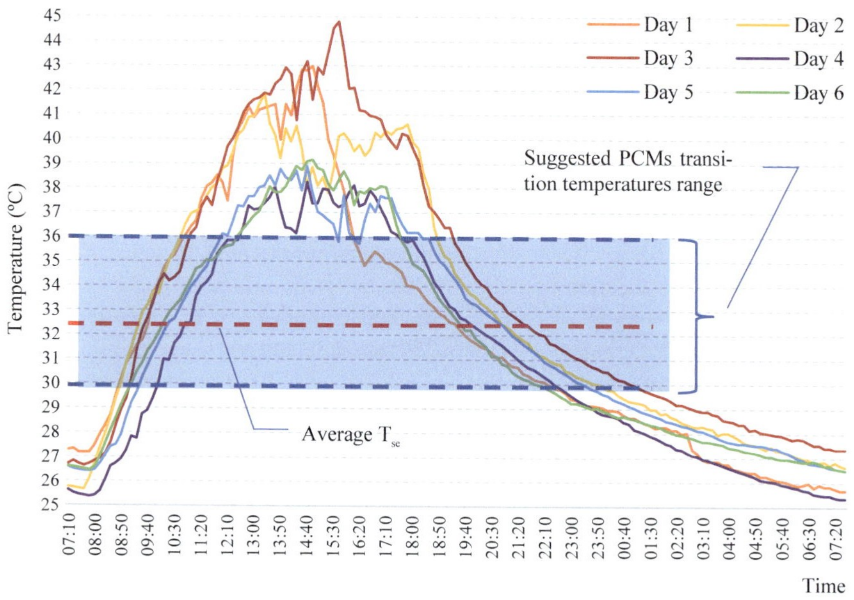

Fig. 8 Suggested PCMs' transition temperature range for the application in buildings' façades based on T_{se} of southern façade during the sunniest days (i.e. days 1–3 at the end of December and days 4–6 at the end of January)

range, the maximum and minimum temperatures must ensure adequate time for the PCMs to fully melt and freeze, respectively. In our case, PCMs melt due to the heat absorbed from the received solar radiation while freeze due to the heat released to the ambient outdoor environment (i.e. PCMs applied on the external surface of building façade). Based on Fig. 8, selecting the transition range between 30 and 36 °C seems to be an adequate one. This means, the selected PCMs must have a melting temperature below 36 °C and freezing temperature above 30 °C to achieve the optimum performance based on the selected days. However, it is still advised to carry out simulation investigation to find out the optimum melting and freezing temperatures for PCMs from the suggested transition temperature range based on their own properties, such as enthalpy of fusion and specific heat capacity as well as PCM quantity.

4 Conclusion

Effects of direct solar radiation on overheating of building façade were investigated in this work. The southern façade that was exposed to the direct solar radiation reached high peaks of T_{se} and T_{si} (i.e. up to 45 °C and 40 °C, respectively) compared to the northern façade that was not exposed to the direct solar radiation (i.e. up to

33 °C and 32 °C for T_{se} and T_{si}, respectively). In addition, the daily fluctuation of T_{se} reached up to 18 °C for the southern façade compared to only 6 °C for the northern facade. This resulted in high peaks of T_{ai} to reach up to 33.73 °C compared to 30.32 °C for southern and northern façades, respectively. In addition, the heat flux through the southern façade was 7.4 times higher than the northern façade. The current study reflects the extreme effects of the direct solar radiations on overheating the building façade, which suggests the importance of implementing passive strategies to tackle these negative effects. Furthermore, this work concluded that there is a potential for the application of PCMs on the external surface of building façades to reduce the overheating. For the investigated conditions, it is suggested to use PCMs with transition temperatures ranged between 30 and 36 °C to ensure complete daily melting/freezing cycle. However, it is still advised to investigate the optimum transition temperature of the PCMs based on thier properties within the suggested range through simulation programs.

Acknowledgements The authors would like to thank the Department of Architecture, Faculty of Engineering, Sana'a University, and the School of Housing, Building and Planning, Universiti Sains Malaysia for the financial support through Research University Grant (1001/PPBGN/814285).

References

1. Dyball DL (2013) An innovative wall-lining for buildings incorporating phase change materials. The University of Brighton in collaboration with Omnova Wallcoverings (UK) Ltd
2. Al-Absi ZAAS, Isa MHM, Ismail M (2019) Application of Phase Change Materials (PCMs) in building walls: a review. In: Awang M, Isa MH (eds) The advances in civil engineering materials. ICACE 2018. Lecture notes in civil engineering, vol 19. Springer Singapore, Singapore, pp 73–82. https://doi.org/10.1007/978-981-13-2511-3_9
3. Akeiber H, Nejat P, Majid MZA et al (2016) A review on phase change material (PCM) for sustainable passive cooling in building envelopes. Renew Sustain Energy Rev 60:1470–1497. https://doi.org/10.1016/j.rser.2016.03.036
4. Bradshaw V (2006) The building environment: active and passive control systems, 3rd edn. Wiley, Hoboken
5. Barzin R, Chen JJJ, Young BR, Farid MM (2015) Application of PCM energy storage in combination with night ventilation for space cooling. Appl Energy 158:412–421. https://doi.org/10.1016/j.apenergy.2015.08.088
6. Tomlinson JJ, Kannberg LD (1990) Thermal energy storage. Mech Eng 41:393. https://doi.org/10.1016/0038-092X(88)90036-9
7. Khadiran T, Hussein MZ, Zainal Z, Rusli R (2016) Advanced energy storage materials for building applications and their thermal performance characterization: a review. Renew Sustain Energy Rev 57:916–928. https://doi.org/10.1016/j.rser.2015.12.081
8. Zhu N, Liu F, Liu P et al (2016) Energy saving potential of a novel phase change material wallboard in typical climate regions of China. Energy Build 128:360–369. https://doi.org/10.1016/j.enbuild.2016.06.093
9. Lei J, Yang J, Yang EH (2016) Energy performance of building envelopes integrated with phase change materials for cooling load reduction in tropical Singapore. Appl Energy 162:207–217. https://doi.org/10.1016/j.apenergy.2015.10.031
10. Engineers daily (2019) Heat transfer in building elements. http://www.engineersdaily.com/2017/10/heat-transfer-in-building-elements.html. Accessed 23 Mar 2019

11. ASHRAE (2004) ANSI/ASHRAE STANDARD 55-2004 thermal environmental conditions for human occupancy. American Society of Heating, Refrigerating and Air-Conditioning Engineers Inc, Atlanta, GA
12. Wahid MA, Hosseini SE, Hussen HM et al (2017) An overview of phase change materials for construction architecture thermal management in hot and dry climate region. Appl Therm Eng 112:1240–1259. https://doi.org/10.1016/j.applthermaleng.2016.07.032

Evaluation of Federal Highway Administration's Traffic Noise Model for Pavements in Qatar

Okan Sirin, Md. Ohiduzzaman and Emad Kassem

Abstract Traffic noise is an irritating problem especially for residents near major transportation corridors. It's in the interest of transportation agencies to check the noise level to develop measures to reduce noise if it exceeds the allowable threshold. This process requires measuring noise level along roadways nearby the residential areas. Alternatively, computer models can be used to study and analyse the noise level which save a lot of resources and enable transportation agencies to take countermeasures when planning new roadways near residential areas. The Federal Highway Traffic Noise Model (TNM) is a computer program that is used to analyse and model traffic noise. The vehicle speed, traffic volume and road geometry are used to predict noise propagation in open far-field environment and determine noise attenuation from noise barriers. In TNMv2.5, pavement and ground type is included but it does not account for noise propagation medium that is random and continuously evolving. In this paper, Federal Highway Administration's (FHWA) TNMv2.5 was evaluated to determine the accuracy of this model in the State of Qatar. During noise measurement, traffic volume and terrain geometry were used to predict the noise levels by using TNMv2.5. The predicted noise levels were compared with the observed levels for checking the accuracy of this model for predicting the future noise levels of roadways. The predicted noise levels were compared with noise levels measured at 19 locations at nine sites throughout the State of Qatar. For typical pavements in Qatar, the TNMv2.5 is overpredicting approximately 2.3 dB for high traffic, whereas it is underpredicting on an average 1.9 dB for low traffic volume. Although the number of test locations is relatively small, it was still established that the results obtained by FHWA's TNMv2.5 model were within the reasonable range of the observed noise levels for pavements in the State of Qatar. This demonstrated that the TNMv2.5 model was suitable for predicting noise level for pavements in Qatar.

O. Sirin (✉)
Department of Civil and Architectural Engineering, Qatar University, P. O. Box 2713, Doha, Qatar
e-mail: okansirin@qu.edu.qa

Md. Ohiduzzaman
WSP, Canberra ACT 2600, Australia

E. Kassem
Department of Civil and Environmental Engineering, University of Idaho, Moscow, ID 83844, USA

M. Awang and M. R. Meor M Fared (eds.), *ICACE 2019*,
Lecture Notes in Civil Engineering 59, https://doi.org/10.1007/978-981-15-1193-6_16

Keywords Wayside noise measurement · Traffic noise model (TNM) · Sound level meter · Asphalt concrete · Pavement

1 Introduction

In the last few decades, traffic volume increased a lot on the highways in order to meet the transportation demand. Thus, more people are exposed to the traffic noise. Therefore, reducing traffic noise is a growing and pressing demand from the people, especially from urban dwellers, where major transportation corridors pass near their houses [1–3]. Transportation authorities worldwide generally predict the noise by using developed traffic noise model based on source of the noise. Over the last few decades, a number of traffic noise models have been developed for predicting noise level caused by moving vehicles. Most of the early works have focussed on predicting noise level from free-flowing traffic. The results of these studies have been used to compute noise originating from the highway and predicting noise level from a similar type of new road carrying traffic travelling at moderate and high speeds.

Traffic noise model (TNM) is a computer software package developed by FHWA in order to predict noise levels in the vicinity of highways facilities. In FHWA's TNM, traffic noise levels near roadways can be predicted based on individual vehicle noise levels, vehicle volume and speed, observer distance and other correlations. The model allows for the incorporation of single and parallel barriers with options including barrier type (walls and berms) and absorptive surface characteristics, roadways with options such as traffic flow and pavement type, the design of terrain geometry, rows of buildings and areas of dense foliage. Transportation agencies and state highways in the United States have used the TNM model as a tool for predicting noise impact in areas near highways. The TNM is extremely important in the United States because it determines the suitability of noise barriers for federally funded projects [4]. As this software continues to evolve, a number of studies are undertaken to validate the model. In this model, three very general pavement categories are included, but for noise prediction an average pavement is required. Burge et al. [5] evaluated TNMv1.0 by comparing field results of longitudinally ground and transversely tined concrete pavements. The results of their study showed similar relationships for both types of pavements using measured and theoretical values.

A study was conducted to compare the noise level prediction using the TNM with the measured sound levels for both calibrated and uncalibrated data [6]. The site variables were not removed for the uncalibrated data, while for the calibrated data both height and distance from the source were considered. In this study, a reference microphone was placed at a distance of 15 m from the centre of the traffic lane and 1.5 m above the elevation of the roadway. The difference between the predicted noise level and the measured one at the reference microphone was calculated, and a correction factor was applied to predict the sound levels at different microphone locations. The results of this study showed that the predicted noise levels correlated well with measured noise levels for both uncalibrated and calibrated data. The authors stated that the general overprediction for the uncalibrated data of the previous TNM versions is no longer an issue due to improvements made on the emission levels in the current version of the TNM. However, the authors explained that the site biases such as pavement type can still affect the predicted sound level. The authors recommended that the TNM noise prediction be calibrated using sound levels measured in the field. Trevino and Dossey [7] conducted a study to validate TNM sound noise predictions using actual roadside measurements taken on PFC pavements. The researchers compared the experimental results to predict TNM sound values for two different pavement types: average (the default option in the TNM software) and open-graded asphalt concrete (OGAC) pavements. The experimental results showed that the TNM tends to overpredict the sound level by almost 5 and 3 dBA when compared to the average pavement and OGAC pavement sound predictions, respectively. The authors recommended using the OGAC option when predicting the noise level on PFC pavements.

Rochat and Read [8] performed validation study for TNMv2.5 by comparing both calibrated and uncalibrated data. They reported that TNMv2.5 performed extremely well due to improvements applied to the implementation of emission level. The latest version of TNMv3.0 addresses the overprediction of noise for earlier versions of the model [6]. This version included only the three original pavement categories, and again the average pavement is required for impact predictions. Although TNM is used by number of transportation authorities in the world for forecasting of noise for roadways but it is still not validated for traffic noise generated from roadways in Qatar. Therefore, the main objective of this paper is to validate the FHWA's TNM by using measured noise level of roadways in the State of Qatar.

2 Details of Measurement Sites

For the purpose of validation of the TNM model, field noise measurements were collected at 19 individual locations contained in nine sites across the State of Qatar. The main purpose of these measurements is to evaluate the accuracy of the model in predicting the noise levels at the receptor of noise. The locations of these sites are shown in Fig. 1. The description of the measured site location is presented in Table 1. At first, visual survey was performed on the tested site to find a suitable location for the purpose of measurement. The sites were selected to achieve a reasonable geographic representation of the State of Qatar and encompass varying topography and elevation. In addition, these measurement sites were selected to embody some features available within TNM, with features such as acoustically soft and hard ground, various terrain geometries and dense foliage to acquire high-quality data measurement.

Fig. 1 General location of noise analysis site

Table 1 Description of test sites and location of sound level meter

Road name	Site number	Measurement number	Position of sound level meter from the centre line		File name in TNM 2.5 software	Type of ground surrounding pavement
			Horizontal distance (m)	Height (m)		
Rawdat Al Rashed Rd	1	M1	7.5	1.5	RR-1	Loose soil
		M2	15	3.7		
Dukhan Hwy	1	M3	7.5	1.5	DH-1	Hard soil
		M4	15	3.7		
	2	M5	15	3.7	DH-2	Loose soil
Salwa Rd	1	M6	7.5	1.5	SR-1	Hard soil
		M7	15	3.7		
	2	M8	15	3.7	SR-2	
	3	M9	15	3.7	SR-3	
Al Shamal Rd	1	M10	7.5	1.5	AS-1	Loose soil
		M11	15	3.7		Grass
	2	M12	7.5	1.5	AS-2	Hard soil
		M13	15	3.7		
QU Rd	1	M14	7.5	1.5	QU-1	Hard soil
		M15	15	3.7		
	2	M16	30	1.5	QU-2	Loose soil
	3	M17	7.5	1.5	QU-3	Hard soil
Al Jumaliya Rd	1	M18	7.5	1.5	AJ-1	Hard soil
	2	M19	30	1.5	AJ-2	Loose soil

3 Instrumentation

All noise measurements were performed by using a consistent measurement methodology in accordance with ISO 11819-1 standard [9] and AASHTO T 98 [10] at each site. The main component of noise measurement system is sound level meter. In this study, Larson Davis: Model 831 sound level meter was used. The main components of sound level meter are microphone, preamplifier and data processing unit. The sound level meter is connected with microphone via preamplifier. Noise data were collected by using two sound level meters: one was placed at 7.5 m (25 ft) with height of 1.5 m (5 ft) and the other one was placed at 15 m (50 ft) with height of 3.7 m (12 ft) from the centre line of the near travel lane and above the ground surface following the ISO 11819-1 [9] and AASHTO T98 [10] standards. The instrumentation of sound level meter is shown in Fig. 2. Sound level meters were placed at specified locations of each site. The sound level meter was calibrated by using Larson Davies: Model CAL 200 calibrator before and after each noise measurement. Sound level meters were set to measure and record A-weight Leq for 15 min.

Meteorological data, such as wind speed, wind direction, air temperature and relative humidity, were collected by using Kestrel 5500 pocket weather meter as acoustical data are greatly influenced by the meteorological conditions. Acoustical data collected during wind speeds exceeding 5 m/s or 11 mph were eliminated during data analysis. Pavement surface temperature was measured by using Fluke thermometer. A portable radar gun was used to measure the speed of the vehicle. Traffic volumes were counted during noise measurement duration. Traffic categories were divided into 'heavy trucks' (semis and other trucks with three or more axels), 'medium trucks' (light trucks with two axles) and all other vehicles were counted as automobiles.

Fig. 2 Experimental setup for wayside noise measurements

4 TNM Analysis

Noise levels were predicted at each measurement location by using TNMv2.5 software. In setting up TNM run, at first site names were identified and English units were selected. Generally, four pavement types, that is, dense graded asphaltic concrete (DGAC), portland cement concrete (PCC), open-graded asphaltic concrete (OGAC) and 'average' (which is derived from DGAC and PCC data) are contained in TNM software. In this study, first noise prediction in TNM software was conducted by considering 'average pavement' and then the predicted and the measured noises were compared.

Measured and predicted sound levels by TNMv2.5 are plotted in Fig. 3. Each red circle is represented as 15 min data block (15 min LAeq). The orange line represents the linear fit whereas two blue lines represent the 95% confidence range. The solid green line is represented as the perfect agreement between measured and predicted noise data. Data points lying above the green line demonstrate the overprediction of noise level by TNMv2.5 and data points that fall below this line symbolize the underprediction of measured noise level.

Figure 3 shows the measured noise levels, the predicted noise levels using TNMv2.5 model by considering average pavement type and the resulting accuracy of each model at each of the 19 site locations (accuracy being defined as the difference between predicted noise level and the measured level). Noise data were measured at both high and low traffic volumes. The comparison of predicted noise levels to measured levels shows that TNMv2.5 is overpredicting approximately 2.5 dB on average for high traffic volume road. However, at low traffic volume pavement, TNMv2.5 is underpredicting approximately 2 dB on average.

All the pavements of the measured sites are DGAC in most of the pavements in Qatar. Hence, noise prediction was again conducted in TNMv2.5 by using 'DGAC pavement'. The direct comparison between TNMv2.5 predicted sound levels and measured data for DGAC pavements in the State of Qatar is presented in Fig. 4.

It can be seen from Fig. 4 that the difference between measured and predicted noise data decreases slightly by using DGAC pavement in TNMv2.5. For high traffic road,

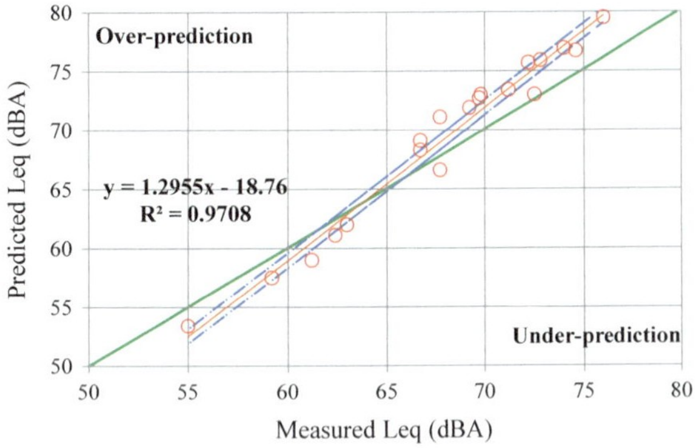

Fig. 3 Measured and predicted noise levels (considering average pavement in TNM software) for pavements in Qatar

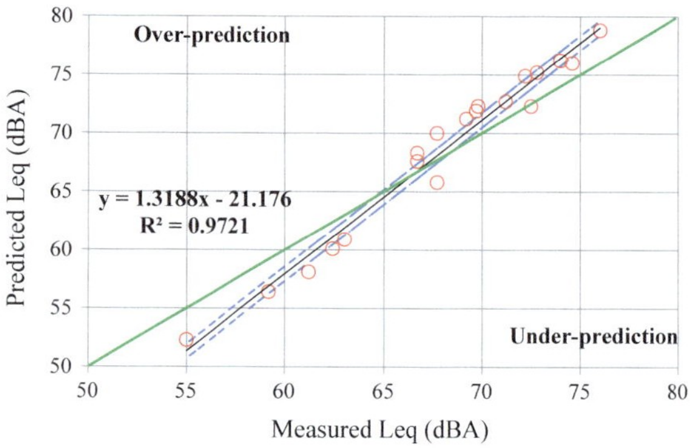

Fig. 4 Measured and predicted noise level (considering DGAC pavement in TNM software) for pavements in Qatar

the TNMv2.5 is overpredicting approximately 2.3 dB, whereas it is underpredicting on an average of 1.9 dB for low traffic volume. However, TNMv2.5 is validated by using 19 sites and site bias is not considered for analysis due to nature of ground but it still predicted noise level within reasonable range. Researchers will further investigate the effect of nature of ground on the performance of TNMv2.5 model for the roadways in the State of Qatar.

5 Conclusions

Many transportation authorities, especially in the United States, use TNM model to predict the noise level of roadways. In this study, the performance of TNM model developed in the United States is evaluated for the prediction of noise level of roadways in the State of Qatar as the roadways are subjected to hot climatic conditions. The comparison between predicted noise levels and measured levels showed that TNMv2.5 is accurate within approximately 2.5 dB on average and statistical basis for high traffic volume, and for low traffic volume, it is accurate within 2.0 dB. This study showed that TNMv2.5 can be used for the prediction of noise level of roadways in Qatar because it is evaluated by using data collected from 19 test sites. The results of this study will be useful for engineers in Qatar to predict the noise level exposure when designing new roads without the need for physical noise exposure measurements, saving time and resources.

Acknowledgements This paper was made possible by the NPRP grant (NPRP 7-110-2-056) from the Qatar National Research Fund (a member of Qatar Foundation). The findings herein reflect the work, and are solely the responsibility, of the authors.

References

1. Sandberg U, Ejsmont J (2002) Tire-road noise reference book. Kisa, Sweden, Informex
2. Ohiduzzaman M, Sirin O, Kassem E, Rochat J (2016) State-of-the-art review on sustainable design and construction of quieter pavements—part 1: traffic noise measurement and abatement techniques. Sustain (MDPI) 8(8):742
3. Sirin O (2016) State-of-the-art review on sustainable design and construction of quieter pavements—part 2: factors affecting tire-pavement noise and prediction models. Sustain (MDPI) 8(7):692
4. Trevino M, Dossey T (2009) Noise measurements of highway pavements in texas. center for transportation research, The University of Texas at Austin, Austin, Texas; Research Report No. FHWA/TX-10/0-5185-3
5. Burge P, Travis K, Rado Z (2002) Transverse-tined and longitudinal diamond-ground texturing for newly constructed concrete pavement: a comparison. Transp Res Rec J Transp Res Board (1792)
6. Rochat J, Fleming G (2004) Addendum to validation of FHWA's Traffic Noise Model® (TNM): phase 1 Report No. FHWA-EP-02-031 addendum and DOT-VNTSC-FHWA-02-01 addendum, U.S. Department of Transportation, Volpe National Transportation Systems Center, Massachusetts
7. Rochat J, Hastings A, Read D, Lau M (2012) FHWA traffic noise model (TNM) pavement effects implementation study: progress report 1. U.S. Department of Transportation, Federal Highway Administration, Washington, D.C.; Research Report No. DOTVNTSC- FHWA-12-01
8. Rochat J, Read D (2013) Effective flow resistivity of highway pavements. J Acoust Soc Am 134(6)
9. International Standard ISO 11819-1 (2005) Acoustics-method for measuring the influence of road surfaces on traffic noise. part 1: the statistical pass-by method. International Organization for Standardization: Geneva, Switzerland
10. AASHTO TP 98-15 (2015) Standard method of test for determining the influence of road surfaces on vehicle noise using the statistical isolated pass-by (SIP) method. American Association of State and Highway Transportation Officials, Washington, D.C

Performance of Steel Slag as a Partial Replacement Fine Aggregate Incorporating Superplasticizer

Gunalaan Vasudevan

Abstract Steel slag is a by-product which can be obtained from steel industry and generated as a waste during the production of steel. Besides, it has been declared a useful construction material, not an industrial waste by most of the developed countries. However, in Malaysia, the rate of utilization of steel slag is very low compared to other developed countries. This paper aims to experimentally study and investigate the effect of steel slag as partial replacement of fine aggregate incorporating superplasticizer in concrete. For this purpose, tested samples consisting of 150 mm concrete cubes and cylinder samples with 150 mm diameter and 300 mm height are prepared and mixed with superplasticizer, namely Sika Visco Crete(VC 2192). Compressive strength, split tensile strength, workability, and ultrasonic pulse velocity of concrete with replacement of fine aggregate by steel slag aggregate are examined and compared with control mix design without replacement of steel slag aggregate. The optimum percentage of replacement of fine and coarse aggregate by steel slag is found. Workability of concrete gradually increases, as the percentage of replacement increases. Yet when beyond 20% of replacement, it starts to decrease, which is found by using slump test. Moreover, compressive strength, tensile strength, and ultrasonic pulse velocity are experimentally investigated. The results point out that for conventional concrete, the partial replacement of fine aggregates by steel slag improves the compressive, ultrasonic pulse velocity, and tensile strength.

1 Introduction

Sustainability is the primary focus in recent years due to the fact that nowadays raw materials are getting reduced and therefore, the use of sustainable materials has been investigated for environmental, economical, and social benefits to improve our quality of life. Concrete is the most commonly used material in construction. Hence the bigger challenge for the future construction firms is to meet the global demand

G. Vasudevan (✉)
Department of Construction Management, Faculty of Built Environment, Tunku Abdul Rahman University College, Kuala Lumpur, Malaysia
e-mail: gunalaanv@tarc.edu.my

M. Awang and M. R. Meor M Fared (eds.), *ICACE 2019*,
Lecture Notes in Civil Engineering 59, https://doi.org/10.1007/978-981-15-1193-6_17

of concrete and find a suitable alternative to natural aggregate for preparing concrete with high compressive strength material [1]. We are also familiar with using steel slag, where steel slag is used as a partial substitute for natural aggregate in the production of asphalt concrete. The use of steel slag from Iron Works Company in combination with natural aggregate as the road base layer and their strength parameters are described in this study. The presence of steel slag depends on the percentage of use and the number of other parameters. This is because almost three quarters of volume of concrete is composed of aggregates and it is becoming a more challenging task to find a suitable alternative material to natural aggregates for preparing concrete. The continuous use of natural rocks will lead to reduction of river beds, which results in the ecological and environmental imbalance. Moreover, nature aggregate will become costly [2]. So to overcome this problem we have to use waste material such as steel slag as a replacement for natural rock or aggregate in the concrete to improve the quality of our living environment.

2 Literature Review

2.1 Steel Slag

Extraction of 'iron' or 'steel' from ores is a very complex process which needed some others materials such as catalysts or flux. After making steel these ingredients forming a matrix are to be periodically weathered up. Steel slag is generated as a by-product of iron and steel-making processes and is produced during the separation of the molten steel from impurities in steel-making furnaces. Moreover, steel slag can be divided into blast furnace slag and steel-making slag. Steel slag occurs when the molten liquid melts and is a complex of silicates and oxides that solidifies upon cooling [3]. According to [4] it is composed of a chemical combination of manganese, silicon, phosphorus, iron, lime. The amount of change in each component is based on the grade of steel being produced and the process used. There are some processes that can affect the range of each compound which are basic oxygen furnace (BOF) and electric arc furnace (EAF) for carbon steel and alloy [5]. Furthermore, the most commonly found steel slag in Malaysia is EAF slag. Electric arc furnace (EAF) process does not use hot metal but cold steel scraps. The EAF method uses high-voltage current to melt the scrap and involves the removal of excess carbon, silicon, and other elements from the molten iron. The electricity has no electrochemical effect on the metal, making it perfectly suitable for melting scrap. During the melting process, other metals including aluminium, silicon, manganese, phosphorus, carbon, and iron are added to the steel to give the required chemical composition. At the same time, oxygen is blown into the EAF to purify the steel. This slag, which floats on the surface of molten steel, is then poured off. The basic oxygen furnace (BOF) process is totally different from the EAF process . In the BOF process, the hot liquid metal from the blast furnace, which contains lime (CaO) and dolomitic lime, is charged into a furnace [3]. A lance is lowered into the converter and then oxygen is injected with high pressure.

2.2 Utilization of Slag

Sustainability materials are the primary focus in recent years. Besides, natural aggregate resources are becoming more difficult to develop or to be removed from the ground. Steel slag can be used as a substitute for natural raw materials like aggregate, which can reduce waste and conserve resources. It protects and preserves our environment and our quality of life. According to [6] steel slag aggregate has good resistance to permanent deformation, high stability with good flow properties, and high stiffness modulus. Moreover, steel slag compared to natural aggregates is lower in costs and performs well when used as a base course, provided it was moist cured to reduce its expansion potential. Benefit from technical advantages offered by many of the steel making slags. High performance products not necessarily low grade applications.

2.3 Steel Slag Properties

Owing to different properties of various aggregates, their level of performance and suitability for an application has been affected. The physical and mechanical characteristic of steel aggregate is the most important part for providing ideal durability, stability, and resistance against abrasion and permanent deformation [7]. Besides, the chemical composition of the steel aggregate was found to be an affecting factor for its adhesion with other construction material to form an ideal combination.

3 Methodology

This research aims to determine the performance of the concrete that contains the steel slag by preparing concrete cube samples and testing them to obtain some of the basic engineering properties. The concrete mix design is done by systematic analysis with the knowledge to choose the proportions of the ingredients to be used to produce an economical concrete mix with the desired strength when hardened. The variables which can be controlled are water/cement ratio, maximum aggregate size, aggregate grading, and the use of admixture (Table 1).

Table 1 Conceptual framework

Phase 1	To identify steel slag as an addition to concrete in percentage	Remarks
Material selection		If no, back to material
	Identify tests/experiments to be carried out	Selection
	Analysis of particle size distribution and properties of steel slag	
Phase 2	Mixture of materials and percentage of steel slag as an addition	If no, back to analysis
Design and conduct		
	Mixing, casting, and compaction	
	Dry, shrinkage, concrete	
	Testing of hardened samples	
Phase 3		
Long-term behaviours	Durability test	
Phase 4	Application (Beam or Column)	
Performance		
	End	

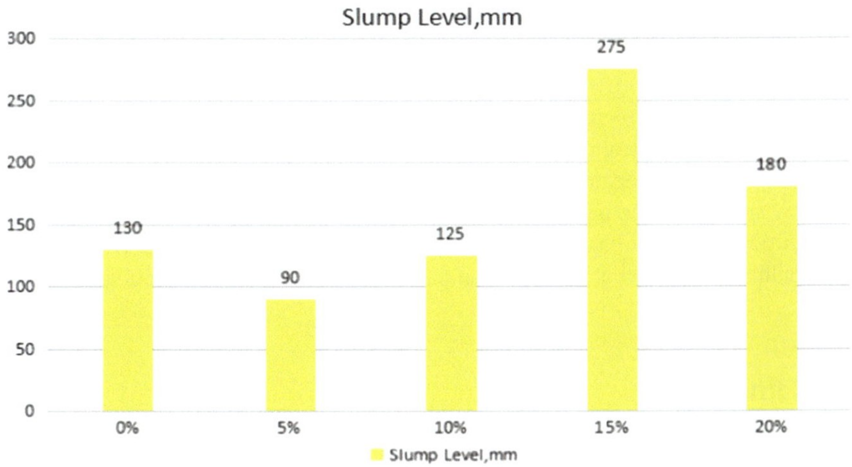

Fig. 1 Workability in terms of slump cone test

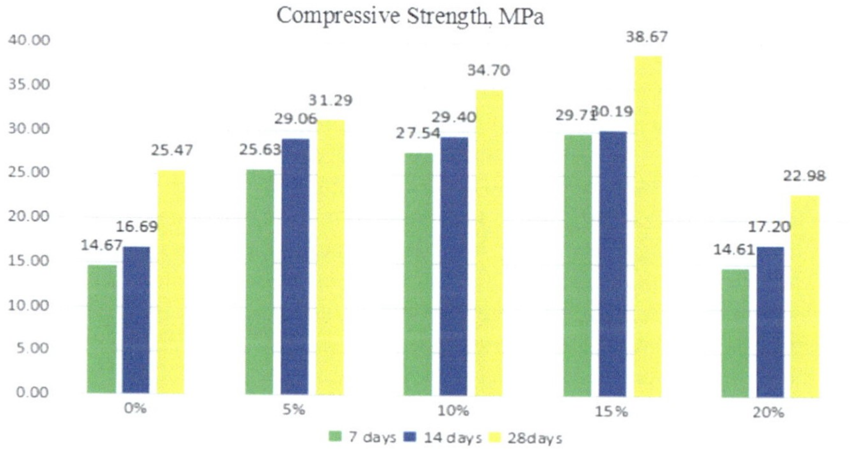

Fig. 2 Compressive strength of steel slag concrete at different curing age

4 Conclusions

4.1 Slump Test

From Fig. 1, it has been shown that normal concrete mix with VC2192, also known as Sika Visco Crete superplasticizer, is giving a slump value of 130 mm. When fine aggregate is replaced by 5% of steel slag, the value of slump drops to 90 mm but the slump value starts to increase if replaced by 10% of steel slag and it reached the highest point of 275 mm. If replaced by 15% the best slump value is obtained in this test. However, once the percentage of the steel slag replacement reaches 20%, the slump starts to decrease, and the decreasing continues as the percentage replacement level increases from 20%. According to [8] it has been shown that river sand is generally finer than that of steel slag. The water absorbing property of steel slag is higher than the normal fine aggregate, so the higher the percentage of replacement of fine aggregates by steel slag, the lesser the workability of concrete (Fig. 2).

4.2 Compressive Strength Test

Qualitatively, on the seventh day of curing, it was observed that concrete increases in the percentage of steel slag replacement to natural fine aggregate. The controlled mix design without replacement of steel slag is at 14.67 MPa, which when compared with the one in which 5% of the aggregate was replaced by steel slag the compressive value is less than 10.96 MPa. Besides, the compressive strength is increased gradually from 0 to 15% and achieved the maximum value at a replacement of 15% on the seventh day of curing with 29.71 MPa, which was different from the controlled mix design with the compressive strength of more than 15.04 MPa. On the 14th day of curing, the compressive strength of controlled mix design increases to 16.69 MPa, which was higher than the seventh day of curing. This proves that the longer the curing time, the higher the strength of the concrete. Moreover, samples with 5, 10, and 15% of steel slag replacement have almost the same compressive strength of 29.06, 29.40, and 30.19 MPa, respectively, on the 14th day. However, as mentioned earlier, on the seventh day of curing when replaced by 20% of steel slag the compressive strength starts to decrease to the value of 17.20 MPa. To summarize, 15% replacement of steel slag had achieved the highest compressive strength compared with other percentage replacements of natural fine aggregate. This is because steel slag concrete is stronger, and the bond between aggregate and paste is more sturdy due to higher angularity and rough surface texture of steel slag compared with fine aggregate.

4.3 Flexural Tensile Strength Test

The results on the 7th, 14th, and 28th day of tensile strength are shown in Fig. 3. From the obtained results, it is observed that the tensile strength values at the end of 7th day increase on 15% of replacement and decrease when up to 20% of fine aggregate replacement of steel slag. This points out that the influence of slag aggregate in resisting tensile strength is significant at only lower values of replacement, beyond which the tensile strength deteriorates. Almost similar behaviour is observed on the 14th and 28th day results. Besides, from the results above, it is obvious that the tensile strength of concrete cast using 15% of steel slag replacement had achieved the best split tensile strength. However, the decrease in tensile strength when the replacement percentage is 20% is due to the high porosity of concrete caused by the increasing

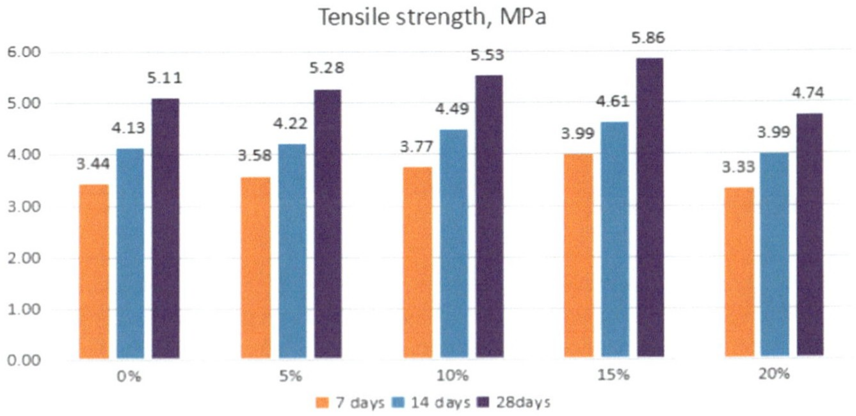

Fig. 3 Tensile strength of steel slag concrete at different curing ages

volume of steel slag aggregates and higher water/cement ratio values. Both factors have reduced the bonding between aggregate and cement mortar paste up to 20% of slag replacement.

4.4 *Ultrasonic Pulse Velocity (UPV) Direct and Semi-direct Test*

From Fig. 4, it was observed that on the seventh day of curing, the sample with 15% of replacement of fine aggregate by steel slag specimens was achieving the highest ultrasonic pulse velocity value in direct test compared to other replacement percentage on the seventh day, which is 31.73 μs. On the 14th day also we observe almost the same data as the 7th day, but have higher and better result data due to longer curing. It was observed from the figure that for the given data in direct test the pulse wave velocity is 34.00 μs on 14th day with 15% of the replacement, which compared with control mix design has increased 17% and achieved the best value in direct test of UPV on 28th day. Owing to this philosophy, the longer the curing time, the better the quality of the concrete. From the results above it can be clearly seen that with the control mix design for direct test, the UPV of concrete is at 29.27 μs on 28th day which is 4.04 μs higher than on seventh day of curing which is only at 25.23 μs. Moreover, the results show that by replacing fine aggregate with steel slag the density of the concrete will be affected.

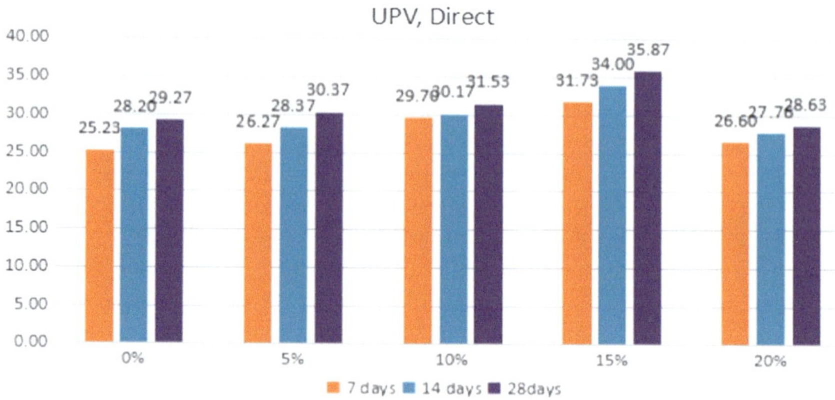

Fig. 4 Ultrasonic Pulse Velocity (UPV) direct and semi-direct test

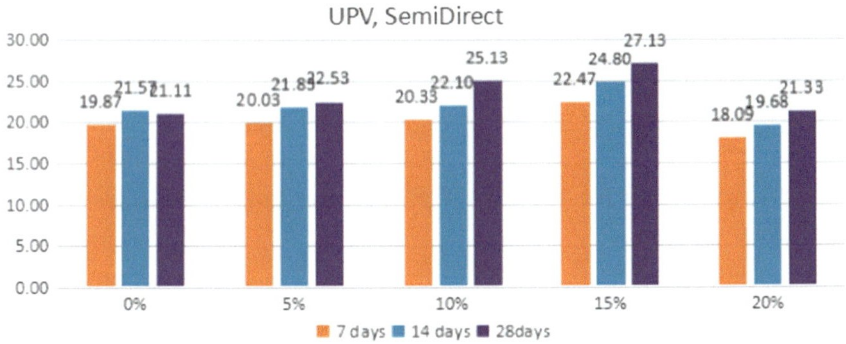

Fig. 5 UPV semi-direct test of steel slag concrete at different curing age

4.5 UPV Semi-direct Test of Steel Slag Concrete at Different Curing Ages

Figure 5 shows the UPV of semi-direct at different ages of curing. From the results it can be clearly seen that semi-direct gives much lower data than direct test. This is because direct test of UPV is more accurate than semi-direct, so the data collected have some difference and shows slower results compared with direct test. In direct test, the optimum level was achieved in 15% of steel slag concrete and the same situation was also obtained in semi-direct test which reached 22.47 μs at the age of 7 days, 24.80 μs at the age of 14 days, and 27.13 μs at the age of 28 days. Moreover, the result was decreased when replacement of steel slag was up to 20% which was the same as direct test at the age of 7th day, 14th day, and 28th day. This analysis shows that the accuracy of ultrasonic pulse velocity test is affected by the concrete age. Apart from that, the comparison between semi-direct method and direct method shows that although direct method shows convenient and satisfactory results upon sensitivity for determining the location of the defect, but the ability to determine the depth of the concrete slab is not possible and it is also not suitable to be used in most situations since it requires access of two surfaces.

5 Conclusions

This part will summarize and conclude the finding of this study. In the slump test steel slag concrete mix has achieved better workability than control concrete mix design. Besides, workability test also shows that replacement of fine aggregate with steel slag in concrete has improved the workability of the concrete [9]. In this test, 15% replacement of fine aggregate has given the best result. However, in the compressive strength test, replacement of fine aggregate with steel slag has given better quality of result, and the optimum percentage of replacement for fine aggregate is 15%, which achieved the best result than other percentages at all ages, beyond which the compressive strength decreases on further replacement. Thorough investigation of results has indicated that split tensile strength of steel slag concrete increases in 14.7% by replacement of 15% of steel slag compared to the conventional fine aggregate concrete at age 28 days. Besides, 15% of replacement has obtained the highest strength than other percentages. Ultrasonic pulse velocity test for steel slag concrete also proves that steel slag replacement of fine aggregate gives better strength and results compared with the concrete without replacement of fine aggregate. Moreover, 15% of replacement gives the highest reading for direct and semi-direct test, which are 35.87 and 27.13 μs, respectively, on 28th day. However, replacement of more than 15% will result in decrease in the reading.

6 Recommendations

The effect of steel slag as a partial replacement of fine aggregate by adding superplasticizer due to various factors is identified in this study. The recommendations to improve the steel slag concrete are as follows:

- Further investigation is needed on the different superplasticizer used together with steel slag in the concrete mix design.
- Further study is required for steel slag mix with other by-products in the concrete mix design.
- Further investigation of the higher percentage of replacement of the fine aggregate by the steel slag aggregate in concrete mix design is required.
- Longer curing time is required in order to understand the behaviour of the concrete, and it might be giving the different result for steel slag concrete.
- Durability test such as sulphate attack test must be further studied in order to understand the percentage loss in weight of the steel slag concrete by sulphate attack.

References

1. Technology FH (2016) User guidelines for waste and byproduct materials in pavement construction. Federal Highway Administration
2. Zumrawi MME, Khalill FOA (2017) Experimental study of steel slag used as aggregate in asphalt mixture. Am J Constr Build Mater 2(1):26–32. https://doi.org/10.11648/j.ajcbm.20170101.12
3. Shi C (2004) Steel slag—its production, processing, characteristics, and cementitious properties. J Mater Civil Eng
4. Borole ST, Shinde RV, Mhaske RB, Pagare SS, Tribhuvan KS, Pawar NM, Tiwari VD, Sanehi AK (2016) Replacement of fine aggregate by steel slag. Int J Innov Res Sci Eng 2
5. Yu X, Tao Z, Song TY (2014) Material properties of concrete with different types of aggregates. In: 23rd Australasian conference on the mechanics of structures and materials

6. Fistri M, Strineka A, Roskovic R (2012) Properties of steel slag aggregate and steel slag asphlat concrete. Institute IGH, Laboratory IGH, Zagreb, Croatia
7. Thangaselvi K (2015) Strength and durability of concrete using steel slag as a partial replacement of coarse aggregate in concrete. Int J Adv Res Trends Eng Technol (IJARTET)
8. Krishna NK, Sandeep S (2016) Study on concrete with partial replacement of cement by rice husk ash. In: IOP conference series: materials science and engineering, vol 149 (2016), p 012109. https://doi.org/10.1088/1757-899x/149/1/012109
9. Quan H-Z, Kasami H (2014) Experimental study on durability improvement of fly ash concrete with durability improving admixture. Sci World J

New Regional-Specific Urban Runoff Prediction Model of Sungai Kayu Ara Catchment in Malaysia

Lloyd Ling, Ming Fai Chow, Wei Lun Tan, Wen Jia Tan, Cheng Yuan Tan and Zulkifli Yusop

Abstract Malaysian government agencies adopted curve number (*CN*) rainfall-runoff model for design use like many other commercial software applications, while most researchers have adopted *CN* values from its published handbook from USA. However, there is no regional-specific curve numbers handbook in Malaysia for the rainfall-runoff predictive modelling. This study did not refer to any *CN* value but derived a statistically significant *CN* value with rainfall-runoff events directly. The derived $\lambda = 0.0002$ is statistically significant ($\alpha = 0.01$), while the optimum *CN* value of 92.95 represents the rainfall-runoff characteristic at the Sungai Kayu Ara catchment. The runoff predictive model estimated an averaged flood depth of 7.46 cm from 100 mm rainfall event when the drainage infrastructure fails to drain away the runoff volume effectively. It is recommended to limit the upstream development, while rainwater harvesting, storm water retention, and detention facilities should be constructed to curb the urban flash flooding at the Sungai Kayu Ara catchment.

Keywords Bootstrapping · Curve number · Runoff prediction · Inferential statistics

---

L. Ling (✉) · W. J. Tan · C. Y. Tan
Civil Engineering Department of Universiti Tunku Abdul Rahman, Lee Kong Chian Faculty of Engineering & Science, Centre of Disaster Risk Reduction (CDRR), Jalan Sungai Long, Bandar Sungai Long, 43000 Cheras, Malaysia
e-mail: linglloyd@utar.edu.my

M. F. Chow
Institute of Energy Infrastructure, Universiti Tenaga Nasional, 43000 Kajang, Selangor, Malaysia

W. L. Tan
Department of Mathematical and Actuarial Sciences of Universiti Tunku Abdul Rahman, Lee Kong Chian Faculty of Engineering, Centre for Mathematical Sciences(CMS) & Science, Jalan Sungai Long, Bandar Sungai Long, 43000 Cheras, Malaysia

Z. Yusop
Faculty of Civil Engineering Department, Centre for Environmental Sustainability and Water Security (IPASA), Universiti Teknologi Malaysia, 81310 Skudai, Johor, Malaysia

M. Awang and M. R. Meor M Fared (eds.), *ICACE 2019*,
Lecture Notes in Civil Engineering 59, https://doi.org/10.1007/978-981-15-1193-6_18

1 Introduction

About 97% of the total water demands come from surface runoff in Malaysia [1]. Runoff volume has increased significantly with the fast pace or urbanisation due to rapidly growing anthropogenic activities [2]. These increased runoff volumes eventually discharged through the watershed outlet. However, downstream areas are mostly flat or in low elevation which are unable to receive and drain the large runoff volume and therefore, flooding frequency increased in the downstream of urban watersheds. As such, it is imminent to develop a methodology to derive regional-specific rainfall-runoff model [3, 4] as it plays an important role in planning and management of water resources [5–7].

2 Study Site and Methodology

The soil conservation service curve number (*SCS-CN*) rainfall-runoff model is one of the simplest models to predict runoff amount in hydrology textbook, which the Malaysian government agencies also adopted, but the accuracy of the model had been questioned by past researchers [8–11]. Equation (1) shows the base *SCS-CN* equation. Initial abstraction (I_a) is the multiplication of λ and S.

$$Q = \frac{(P - I_a)^2}{P - I_a + S} \tag{1}$$

where,

P Depth of a rainfall event (mm)
Q Runoff depth from a rainfall event (mm)
S Water retention depth of a catchment (mm)
I_a Initial retention or abstraction depth (mm).

The base *SCS-CN* equation consists of three main parameters, which are P, λ, and S. The equation is simplified by fixing $\lambda = 0.2$, as shown in Eq. (2). Equation (2) has a constraint that P must be larger than I_a, else there will be no runoff [12].

$$Q = \frac{(P-0.2S)^2}{P+0.8S} \tag{2}$$

The relationship of *CN* and *S* is shown in Eq. (3). *CN* has no unit, and its value has spanned from zero to 100 to reflect a catchment's infiltration ability. The *CN* value closer to 100 indicates a catchment with high saturation condition, and therefore, will produce more runoff. As a result, the random *CN* value selection will lead to meaningless runoff prediction results [13]. However, most of the SCS practitioners did not explore site-specific calibration possibility and often used "trial and error" method to identify *CN* value which will match the observed runoff. At the end, the choice of *CN* to represent a catchment becomes non-scientific and inconsistent to a researcher [14].

$$CN = \frac{25,400}{S + 254} \tag{3}$$

where,

CN = Curve number value

The new regional-specific *SCS-CN* model and *CN* derivation technique can solve λ and *CN* problems as aforementioned. Basically, this new calibrated *SCS-CN* model maintains the parameters available in base *SCS-CN* equation. The proposed *CN* derivation methodology in this article used nonlinear numerical optimisation algorithm with non-parametric inferential statistics and BCa procedure of bootstrapping method [15] to derive λ and *S* values from Eq. (1) using the rainfall-runoff (*P-Q*) dataset collected at the Sungai Kayu Ara catchment.

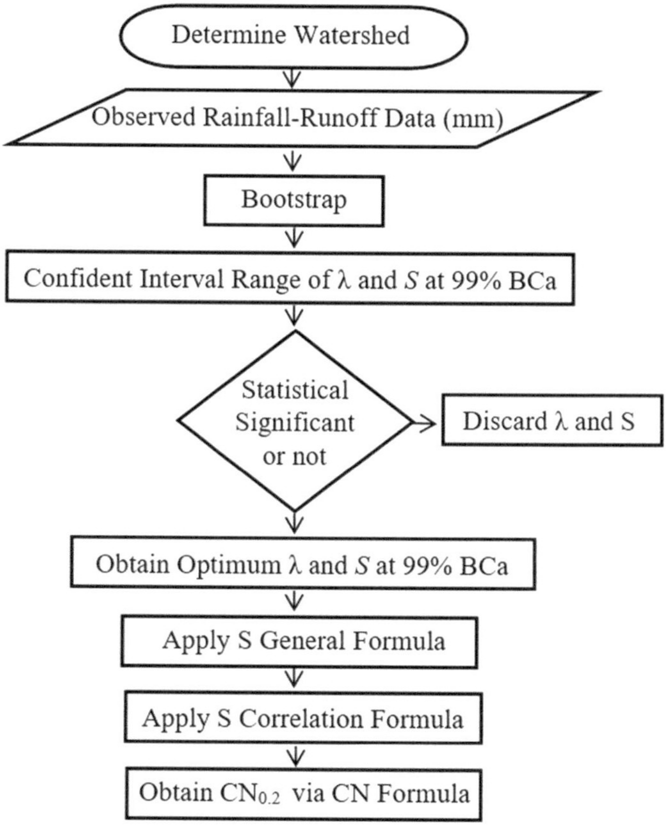

Fig. 1 Methodology flowchart of this study

This study adopted the S general formula (Eq. 4) from Ling and Yusop [4], which solves for S in terms of P, Q, and λ to allow the conversion of new derived λ value back to the mathematical scale, where $\lambda = 0.2$ in order to calculate the conventional CN value, which has been in use since 1954. The obtained $S_{0.2}$ value will be substituted into the CN equation (Eq. 3) to find $CN_{0.2}$ [16]. The overall flowchart of calibrated *SCS-CN* rainfall-runoff model is stated in Fig. 1.

$$S_\lambda = \frac{[P - \frac{(\lambda-1)Q}{2\lambda}] - \sqrt{PQ - P^2 + [P - \frac{(\lambda-1)Q}{2\lambda}]^2}}{\lambda} \tag{4}$$

Sungai Kayu Ara catchment is located with N 30 6′ to N 30 12′ and E 1010 35′ to E 1010 39′ in Damansara, Kuala Lumpur, which covers an area of 22.33 km^2, as shown in Fig. 2. The river originates from a forest reserve in the upstream of Sungai Penchala and flows towards the suburban areas of Desa Sakti, Kampung Sungai Penchala, and Taman Tun Dr. Ismail. The river flows to the relatively flatten developed area (WL 4, lower left dash circle in Fig. 2) near to the confluence of Sungai Damansara [17]. There are 93 storm events between 1996 and 2006 in this study. The lowest recorded rainfall depth was 1.33 mm with measurable runoff amount of 0.56 mm.

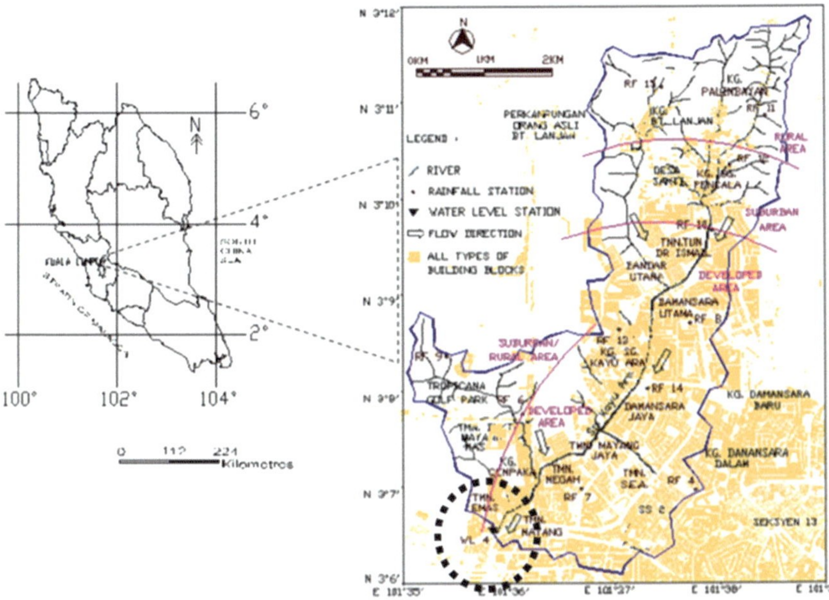

Fig. 2 Location of Sungai Kayu Ara catchment (*Source* Abustan et al. [17])

3 Results and Discussion

Table 1 shows the descriptive statistics of mean and median λ values together with λS inferential statistics (BCa) confidence interval range of 99%. Shapiro–Wilk test result was used to select mean or median λ value as the sample size of this study is less than 2,000 samples. The p value of Shapiro–Wilk test is less than 0.05 for derived λ value; thus, the dataset is not normally distributed, which means the median λ value is more suitable to be used for the runoff predictive model formulation to represent the rainfall-runoff conditions at Sungai Kayu Ara catchment.

The BCa 99% confidence interval range of the standard deviation for λ value indicated that λ cannot be a constant but a variable which spanned from its lower confidence interval limit of 0.0006 to its upper limit of 0.2004. On the other hand, neither the mean nor the λ confidence interval spans include the λ value of 0.20. Thus Eq. (2) becomes statistically invalid to be used to predict runoff at Sungai Kayu Ara catchment.

The descriptive statistics results of Sungai Kayu Ara catchment were tabulated in Table 2. The optimised median λ value is 0.0002, while $S_{0.0002}$ is 33.07 mm for

Table 1 Descriptive statistics of λ at Sungai Kayu Ara catchment

Descriptive statistics	λ	BCa 99% Confidence Interval	
		Lower bound	Upper bound
Mean	0.0113	0.0004	0.0427
Median	0.0002		
Skewness	9.6356	5.3775	9.6357
Kurtosis	92.8939	27.5254	92.8959
Std. deviation	0.1019	0.0006	0.2004
Shapiro–tests of normality			
	Statistics	p value	
λ	0.085	0.000	
S	0.742	0.000	

Table 2 Statistical results of the runoff predictive model

p value	<0.001
E	0.832
BIAS	−0.410
RSS	1,974.874
Optimum λ	0.0002
Optimum S	34.067
Equivalent $CN_{0.2}$	92.95
99% Confidence interval range of equivalent $CN_{0.2}$	[92.95, 94.96]

Table 3 Predicted runoff volume of Sungai Kayu Ara catchment

P (mm)	Q (mm)	Runoff Volume (m^3)	No. of olympic size pools	Averaged flood depth (cm)[a]
70	47.08	1,051, 289.57	421	4.71
80	56.10	1,252, 755.67	501	5.61
90	65.28	1,457, 741.63	583	6.53
100	74.58	1,665, 459.79	666	7.46

[a]Actual flood depth can be deeper at the flat land areas near to the outlet

Sungai Kayu Ara catchment. All the results obtained are statistically significant at alpha = 0.01.

Since $I_a = \lambda S$, substitution of λ and S values produces $I_a = 0.006$ mm. Substitution of I_a and S into Eq. (1), the Sungai Kayu Ara catchment urban runoff prediction model can be formulated as:

$$Q_{0.0002} = \frac{(P-0.006)^2}{P+34.061} \tag{5}$$

where, $Q_{0.0002}$ = Runoff amount (mm) of new model formulated with $\lambda = 0.0002$.

Equation (5) is the calibrated *SCS* urban runoff model given that measurable rainfall depth must be larger than 0.006 mm, or else there will be no measureable runoff. The Sungai Kayu Ara catchment runoff prediction model (Eq. 5) is also statistically significant ($p < 0.001$), while the correlation between $S_{0.0002}$ and $S_{0.2}$ can be determined by using SPSS as:

$$S_{0.2} = 0.823 S_{0.0002}^{0.893} \tag{6}$$

Equation (6) has high adjusted R^2 of 0.977, low standard error value of 0.190, and $p < 0.001$. Subsequently, the equivalent $S_{0.2}$ can be calculated as 19.26 mm leading to the derivation of $CN_{0.2} = 92.95$. Table 3 summarises the runoff volume with respect to several rainfall event scenarios.

4 Conclusions

In conclusion, Eq. (5) can be used to predict the runoff conditions of the Sungai Kayu Ara catchment. The new regional-specific *CN* derivation method used non-linear (supervised) optimisation technique through the guide of inferential statistics and derived *CN* value of 92.95 with 99% confidence interval ranging from 92.95 to 94.96 to represent the Sungai Kayu Ara catchment. When 70 mm rain falls in Sungai Kayu Ara catchment, about 1 million m^3 runoff will be produced (approximate to 421 Olympic size swimming pools). When the rainfall depth reaches 100 mm, the total runoff volume will increase from 1 million to 1.7 million m^3 (equivalent to 666 Olympic size swimming pools). The averaged flood depth throughout the catchment increased from 4.71 to 7.46 cm in the event that the drainage infrastructure fails to drain away the runoff volume effectively. It is recommended to limit the upstream development, while rainwater harvesting, storm water retention, and detention facilities should be constructed to curb the urban flash flooding of this catchment.

Acknowledgements The authors would like to thank Universiti Tunku Abdul Rahman and Brunsfield International Group for their financial support in this study. This study was also supported by Universiti Teknologi Malaysia and Universiti Tenaga Nasional. The authors would also like to acknowledge the guidance from Professor Richard H. Hawkins (University of Arizona, USA) and Professor Ismail Abustan (USM) for sharing his research data.

References

1. Department of Irrigation and Drainage (DID) (2000) Urban stormwater management manual for malaysia. Kuala Lumpur, Malaysia
2. Adams BJ, Papa F (2000) Urban stormwater management planning with analytical probabilistic models. Wiley, New York
3. Hawkins RH, Khojeini A (2000) Initial abstraction and loss in the curve number method. Arizona. Hydrological Society Proceedings
4. Ling L, Yusop Z (2018) Derivation of region-specific curve number for an improved runoff Prediction accuracy. In: Improving flood management, prediction and monitoring (Community, environment and disaster risk management, vol 20). Emerald Publishing Limited, pp. 37 – 48. Chapter 4
5. Ajmal M, Kim TW, (2015) Quantifying excess stormwater using *SCS-CN*-based rainfall runoff models and different curve number determination methods. J Irrig Drain Eng (ASCE) 10(6):1943–4774
6. Banasik K (2009) Predicting the flood hydrographs for small urban watersheds (in Polish). Warsaw University of Life Sciences—SGGW Press, Warsaw
7. Chan CH (2005) Modelling watershed cumulative effects on runoff based on major land uses. MS thesis
8. D'Asaro F, Grillone G, Hawkins RH (2014) Curve number: empirical evaluation and comparison with curve number handbook tables in Sicily. J Hydrol 19(12):1–13
9. Hawkins RH, Jiang R, Woodward DE, Hjelmfelt AT, van Mullem JA, Quan QD (2002) Runoff curve number method: examination of the initial abstraction ratio. In: Proceedings of the second federal interagency hydrologic modeling conference. ASCE Publications, Las Vegas
10. Yuan Y, Nie J, McCutcheon SC, Taguas EV (2014) Initial abstraction and curve numbers for semiarid watersheds in south eastern Arizona. Hydrol Process 28:774–783
11. Zhou SM, Lei TW (2011) Calibration of *SCS-CN* initial abstraction ratio of a typical small watershed in the loess hilly-gully region. Agric Sci China 44:4240–4247
12. NRCS (2001) Engineering field handbook. National engineering handbook, Part 650, Chapter 14, Water management (Drainage)
13. Hawkins RH (1993) Asymptotic determination of runoff curve numbers from data. J Irrig Drain Eng 119(2):334–345
14. Schneider L, McCuen RH (2005) Statistical guidelines for curve number generation. J Irrig Drain Eng 131:282–290
15. Efron B, Tibshirani R (1994) An introduction to the bootstrap. Chapman & Hall/CRC

16. Hawkins RH, Ward T, Woodward DE, Van Mullem J (2009) Curve number hydrology: state of the practice. Reston, Virginia, ASCE
17. Abustan I, Sulaiman AH, Wahid NA (2008) Urban rainfall-runoff study to validate the design chart in the Malaysian urban stormwater management manual (MSMA). In: Online proceeding of 11th international conference on urban drainage, Edinburgh, Scotland, UK

Connector Design Selection for Modular Forest Bridge Using Finite Element Analysis

M. Rizuwan Mamat and W. M. Shukri W. Ahmad

Abstract Forest harvesting in Malaysia is currently being carried out deep inside the forest where accessibility is the main problem. Under the current logging road specification, temporary bridges are built with log stringer for temporary usage and are removed or left to deteriorate at the end of the use period. This situation leads to a problem to whoever needs to cross the stream for official or personal purposes. This road is customarily used only for a short period with a prolonged usage cycle, and this leads the management to close these roads whenever there are no logging activities. Most of the time upon completion of logging activities, these temporary bridges are not maintained and will deteriorate and collapse due to the high velocity of water flow, erosion, and sedimentation. Modular and mobile concepts of the bridge are recommended for easy transportation, installation, and removal for reuse at multiple sites. This modular mobile forest bridge consists of the segmented timber beams which are joined by aluminium connector. Therefore the critical part of this structure is at the joint where failure might occur when the loading is applied. Five connector designs had been recognised to join the segmented timber beams. Finite element analysis (FEA) was used under static stress analysis with five different load cases to carry out the simulation. Modulus of resilience was calculated to determine the capability of the material to store energy under the deflection. Deflection analysis and volume of connectors were also analysed for result comparison.

Keywords FEA · Aluminium · Segmented beam · Girder · Static stress · Timber

1 Introduction

Forest harvesting in Malaysia is currently being carried out deep inside the forest where accessibility is the main problem. Stream or river crossing requires a bridge, and under the current specification, this bridge is built with log stringer for only temporary usage and removed or left to deteriorate at the end of the use period

M. R. Mamat (✉) · W. M. S. W. Ahmad
Forestry and Environmental Division, Forest Research Institute Malaysia (FRIM), 52109 Kepong, Selangor Darul Ehsan, Malaysia
e-mail: rizuwan@frim.gov.my

M. Awang and M. R. Meor M Fared (eds.), *ICACE 2019*,
Lecture Notes in Civil Engineering 59, https://doi.org/10.1007/978-981-15-1193-6_19

[11]. Almost any species of wood would allow usage less than three years. Proper culvert design with hardwood species for the sill log will extend the service life [2]. Forest department (FD) faced this problem to run forest and soil treatment in the logged-over area for post-harvesting activities. Most of the time, upon completion of logging activities, these temporary bridges are not maintained and will deteriorate and collapse due to the high velocity of water flow, erosion, and sedimentation [6]. Therefore, there is a need to have a suitable, cost-effective, and easy to handle but durable bridge that can be used to access the logged-over forest to carry out post-harvesting activities. Taylor et al. [11] suggest a portable bridge to ease transportation, installation, and removal for reuse at multiple sites that would reduce the construction cost compared to a permanent structure, and at the same time, reduces potential water quality problems [1]. Nowadays, there are number of portable bridges developed to reduce the environmental impacts, but still, the installation and dismantling need heavy machinery.

One of the necessary criteria in designing these portable bridges is the consideration of easy assembly, installation, removal, and transportation. In forestry industry, light construction equipment and minimum field cost and work for installation and removal are expected [11]. The ability of portable bridges to serve multiple installations makes them much more economically feasible than a permanent structure In addition, they have the potential to avoid water quality problems [11]. In timber structural components, joints are often the weakest points. Knowledge of the mechanical behaviour of the joints in designing timber structural is critical. A few factors that might affect the performance of the joints are the geometry of the members and joints, the orientation of the load concerning timber grain, and clearance and friction between the joint members [12]. This modular bridge is evaluated as a single system as the durability of each component relies on the durability of other components [3].

2 Materials and Methods

The cross-section of the I-beam is 80 mm × 240 mm, where the flange cross-section is 80 mm × 60 mm. Five connectors were designed based on this cross-section (Fig. 1). The designs of the connector are supposed to support the applied loading and allow the stress transfer from one end to the other end of the timber beam. This analysis mainly focuses on the connector that would be used to join the segmented timber beam. The material used for the connector is aluminium. Static stress analysis was carried out for those designs of connectors. The results generated from the simulation show the performance of the connector to sustain the loading.

The analysis was carried out using simulation of static stress on different connector designs. Five connector designs had been recognised to join the segmented timber beam. These connectors and timber beams were modelled in 3D using AutoCAD® program and were assembled before being transferred to Auto-desk Mechanical Simulation® program. These assembled models were simulated using static

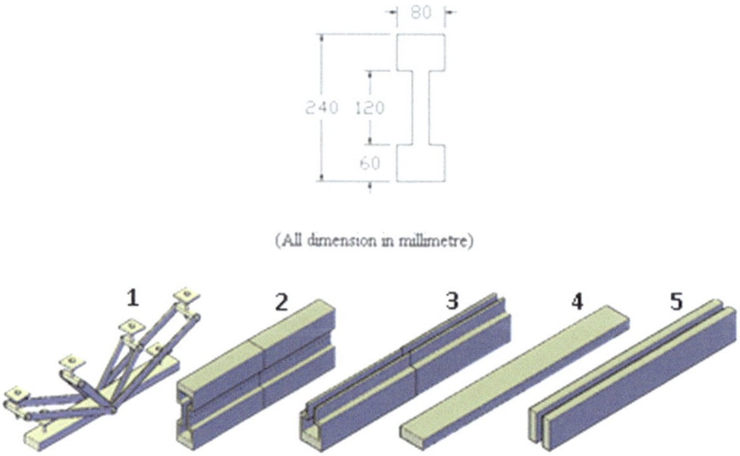

Fig. 1 I-shaped cross-section and connectors design

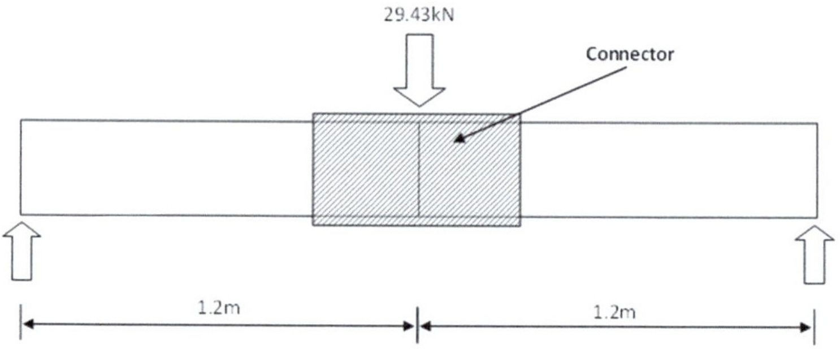

Fig. 2 Specimen assembly

stress analysis to determine the best selection of aluminium connectors for actual fabrication.

Two units of this timber beam were assembled with an aluminium connector before being transferred to finite element analysis for bending test simulation. The simulation was carried out for all connector designs using Autodesk Mechanical Simulation® under static stress analysis. All connected parts in the simulation program were assigned as surface contact type considering the mechanical connection during the assembly. In the FEA, a downward maximum nodal force of 29.43 kN (3000 kg) was assigned. This specified load was based on gross vehicle weight of a Toyota Hilux four-wheel-drive, with a maximum payload of 1060 kg. Five load values were inserted into the simulation: 5.89, 11.77, 17.66, 23.54, and 29.43 kN. The applied load was placed at the beam mid-point (Fig. 2). Stress, strain, deformation, and displacement results were compiled for further analysis.

Table 1 Material properties of timber and aluminium

Material properties	Orthotropic	Isotropic
	Timber	Aluminium
Mass density (kg/m^3)	425.338	2700
Modulus of elasticity (MPa)	11100.559	68900
Shear modulus of elasticity (MPa)	577.091	2600
Poisson's ratio	0.329	0.33

Timber [5], Aluminium [4]

Modelling and assembly of the timber beam and connector 3D models were performed in AutoCAD®. Both timber and aluminium were assigned as brick elements in the simulation program. Material properties that were previously established for Western white pine and aluminium (Table 1) were used as inputs for the FEA simulation.

The developed 3D models were assumed to be geometrically linear which could respond to the system with linear elasticity [7]. The results generated by the static stress analysis were assumed not to be sensitively dependent on the selected mesh size [8].

3 Results

3.1 Stress–Strain Analysis

In the stress–strain analysis, the maximum value of stress and strain generated from the simulation were recorded and analysed. The evaluation was carried out to review the modulus of resilience generated from the stress–strain curve. In the FEA in this section, Tan and Smith [10] convinced that accurate stress distributions in members were obtained in the elastic range of analysis. The comparison using the modulus of resilience, as Roylance [9] states where the material can absorb the energy without suffering damage, shows that the higher the modulus of resilience, the more energy it can store. These values were acquired by calculating the area under the stress–strain curve. Connector 1 recorded the highest resilience (1211), followed by Connector 4, Connector 2, Connector 5, and Connector 3 with resilience values of 371, 40, 18, and 15, respectively. These values show the capability of the connector design to withstand the loading compared with other designs.

3.2 Displacement Analysis

This analysis focused on vertical measurement during deformation specimen when the loading was applied. Each connector reacted and responded with different behaviours under the applied loading. In three points bending test, the highest displacement and deformation occurred at the mid-span of the specimen. In the displacement analysis the design of Connector 1 recorded the highest displacement with 0.045 m, followed by Connector 4, Connector 2, Connector 3, and Connector 5 with displacement values of 0.029, 0.026, 0.025, and 0.022 m, respectively. Displacement analysis was calculated for the whole specimen body. Since this study analyses the hybrid material which was assembled by the connector at the mid-span of the specimen, therefore, displacement result could not be used as the main parameter to determine the connector design.

3.3 Volume Analysis

Volume is considered in the selection connector design because the weight of structure affects the transportation, handling, and assembling times at the installation site. The highest weight of the connector is from Connector 3 followed by Connector 4, Connector 5, Connector 2, and Connector 1 with the volumes of 0.009, 0.008, 0.007, 0.006, and 0.003 m^3, respectively.

4 Discussion and Conclusion

In this specific analysis, the result of modulus of resilience was calculated based on the assembly of segmented timber beam and aluminium connector. Modulus of resilience reflects the capability of the material to store energy under the deflection. Therefore, the selected connector design should be capable of sustaining the stress during the applied loading. Thus, a higher value of modulus of resilience shows higher capability of the connector design to withstand the loading. The weight of the structure, which is reflected from the volume of the connector design, will eventually impact the process of installation and removal of the modular and mobile bridge structure. Thus, a lighter structure is recommended. Connector 1 showed a promising result based on resilience and volume analysis. However, displacement analysis shows that the timber beam is more suggestible for reinforcement to reduce the displacement if Connector 1 is selected. This concept of the modular mobile bridge allows the forest manager to reach the remote areas in the forest by using this kind of bridge for stream crossing, and also, it can be used on steep terrain. These modular and mobile concepts need customised designs and lightweight characteristics of materials. Good modular design with the right selection of material will ease the handling, assembling, installation, and also maintenance purposes.

Acknowledgements The authors gratefully acknowledge the use of the facilities at Forest Research Institute Malaysia (FRIM). This study was funded by project P23085100018005 under the Eleventh Malaysia Plan.

References

1. Brinker RW, Taylor SE (1997) Portable bridges for forest road stream crossings. Alabama Cooperative Extension System, Florence
2. Chow B (2013) Engineering manual. Forest service, British Columbia
3. Dexter RJ, Osberg CB, Mutziger MJ (2001) Design, specification, installation, and maintenance of modular bridge expansion joint systems. J Bridge Eng 6:529–538
4. Dey P, Narasimhan S, Walbridge S (2017) Evaluation of design guidelines for the serviceability assessment of aluminum pedestrian bridges. J Bridge Eng 22:04016109. https://doi.org/10.1061/(ASCE)BE.1943-5592.0000983
5. Forest Products Laboratory (1999) Wood handbook—wood as an engineering material. General Technical Report FPL–GTR–113. U.S. Department of Agriculture Forest Service, Madison
6. Leete R (2008) Malaysia sustainable community: forest management in Sabah
7. Minalu KK (2010) Finite element modelling of skew slab-girder bridges. M.Sc. thesis, Technical University of Delft, Netherlands
8. Park J, Yindeesuk S, Tjhin T, Kuchma D (2010) Automated finite-element-based validation of structures designed by the strut-and-tie method. J Struct Eng 136:203–210
9. Roylance D (2001) Stress-strain curves. https://doi.org/10.1361/aac
10. Tan D, Smith I (1999) Failure in-the-row model for bolted timber connections. J Struct Eng 125:713–718
11. Taylor SE, Ritter MA, Murphy GL (1995) Portable glulam timber bridge design for low-volume forest road. In: Proceedings of the 6th international conference on low-volume roads, vol 2, pp 328–338
12. Xu B, Bouchaïr A, Racher P (2014) Appropriate wood constitutive law for simulation of nonlinear behavior of timber joints. J Mater Civ Eng 26(6):1–7. https://doi.org/10.1061/(ASCE)MT.1943-5533.0000905

Applying Fire Simulation to BIM Modeling with API Programming for Evacuation Time Calculation

Te-Wei Fan

Abstract Revit secondary development (Revit Application Programming Interface, abbreviation Revit API) is currently and often used within Revit, and it is the Revit development trend of future BIM software. BIM is already closely tied to the construction industry, and BIM is inseparable from some large public infrastructure projects. In this literature, the Revit API is being used to derive the parameter data from the Revit model to carry out calculations for analysis and study on the feasibility of using the Revit API to draft the personnel evacuation program.

Keywords BIM · API · Evacuation

1 Introduction

Building Information Modeling is abbreviated as BIM. On the basis of the various related information and data of the construction project, the construction model is being established through simulation of digital information to construct real information. BIM is not just simply the integration of digital information but the digitalized method of using digital information to design, construct, and manage. Such a method to support the integrated management environment of a construction project can enable the construction project to significantly enhance efficiency and greatly reduce risk. This maximizes the reduction of investment in construction costs and achieves the goal of maximizing profits for the investment parties of the construction. Hence, BIM can be said to be the development trend for construction in the future.

T.-W. Fan (✉)
Department of Civil Engineering, Chung Hua University, No. 707, Sec. 2, Wufu Rd., Xiangshan District, Hsinchu City, Taiwan
e-mail: dfan@chu.edu.tw

M. Awang and M. R. Meor M Fared (eds.), *ICACE 2019*,
Lecture Notes in Civil Engineering 59, https://doi.org/10.1007/978-981-15-1193-6_20

So products based on Revit are parametric BIM tools, and these tools are similar to CAD in the purpose of assisting with computing design. They can be used with 3D modeling or 2D diagrams.

The use of API within Revit can help with seamless design of add-on applications that you want and it can be synchronized with Revit user interface and commands. The API can also be used to help us execute the repeated steps within the design, such as capturing of information from the modeling diagram, selection and setting up of single structure, and to carry out analysis on the downloaded Revit model.

2 Literature Review

Wang et al. [1] presented an evaluation method by fire simulation combined with BIM modeling for calculating and assessing the evacuation time required with allowable time permitted for safety evaluation in fire condition. Subsequent efforts showing that how BIM models were implemented to help determine the evacuation planning with the information provided [2].

In the thesis written by Lin [3] from the National Central University, the Fire BIM fire disaster investigation system of the Revit API was used in a Revit model that was already set up to collect and analyze building information after fire disasters. In comparison to the traditional handwritten post-disaster reports, it is more authentic and convenient.

The study by Jia et al. [4] utilized the FDS evacuation software to add to the social force model proposed by the scholar Helbing, and further used the evacuation time formula proposed by Togawa, where analysis was carried out on different density conditions of 80 people, 160 people, and 240 people. Comparisons on the results of different positions and width of escape routes were made. It was concluded that the wider the exit routes, the shorter the evacuation time.

In the study by Chen and Yang [5], Simulex of IES was used in combination with the FDS system. The study was carried out in conjunction with the building regulations of the Taipei Dome, to provide some recommendations for modification of evacuation routes in the Taipei Dome.

In the study by Zhuang [6], the Autodesk Revit modeling was used with the FDS system to simulate the fire disaster process. This similarly proves that the use of BIM can be beneficial to disaster prevention, i.e., evacuation during the occurrence of a fire disaster.

3 Method

Model setup: Upload the Autodesk diagram of 6F in the administration building of Chung Hua University as shown in Fig. 1 and the required fire parameters as shown in Fig. 2 into Autodesk Revit. Following which the pillars, walls, and floor boards will be built according to detailed measurements as shown in Figs. 3 and 4.

Drafting and Execution of Revit API [7], tabs are considered to be added in advance, and two models are separately used to analyze the evacuation conditions of personnel.

1. The same evacuating person is given different escape routes and then the evacuation time is calculated subsequently after the operator has selected the escape route.
2. Preset the position and escape route of the evacuating person according to the different number of people for selection (1–5 persons). The total evacuation time is calculated after the selection is made

The main formula for personnel evacuation analysis is the one proposed by Togawa [8]:

T is the escape time; P is the total number of personnel escaping; d is the width of the exit; N is the exit flow volume; L is the distance of the farthest exit from the room; *v* is the speed of the personnel.

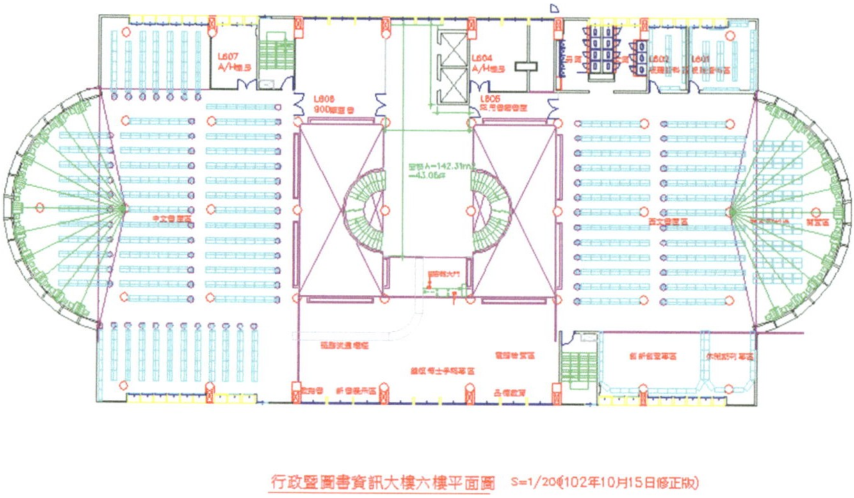

Fig. 1 CAD drawing of the administration building

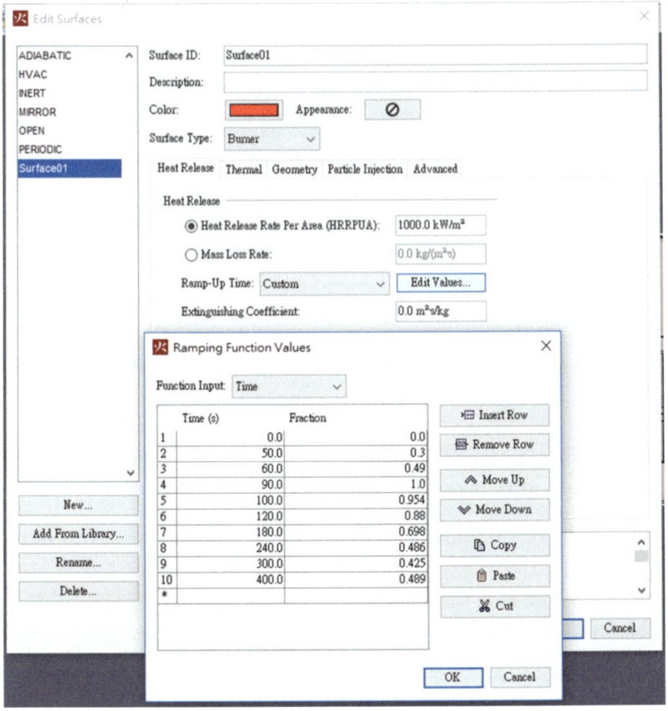

Fig. 2 Fire parameters for modeling

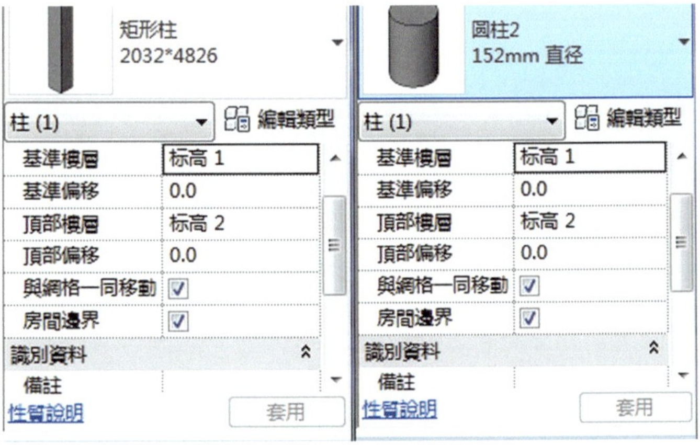

Fig. 3 Dimensions and attributes of columns

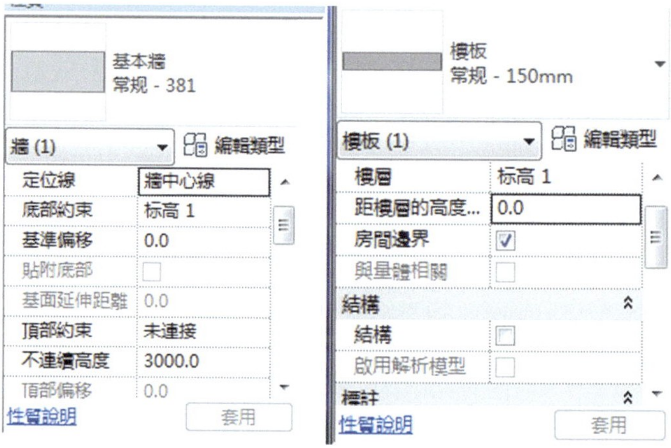

Fig. 4 Dimensions and attributes of floors and walls

The entire model established is shown in Fig. 5.

Assuming its variable v = 5 m/s, and according to the reference information, the age of the evacuating personnel will affect the value of the escape flow volume. The primary range is distributed between 1.25 and 1.97 p/m/s. We select the middle value 1.6 p/m/s and assuming the exit door width is d=3 m. For the distance from L, we select the total length of the preset modeling line at the time of Revit modeling obtained from the Revit API. The escaping personnel P is determined according to the situation.

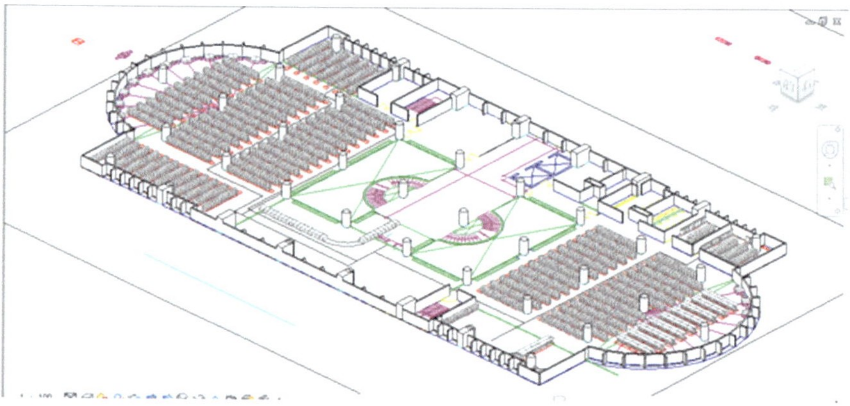

Fig. 5 3D model established by Revit

The Code for panel are listed as follows:

```
using System;
using Autodesk.Revit.UI;
using Autodesk.Revit.DB;
using System.Windows.Media.Imaging;
namespace Evacuation
{
  [Autodesk.Revit.Attributes.Transaction(Autodesk.Revit.Attributes.TransactionMode.Ma
nual)]
     public class CsEvacuation : IExternalApplication
     {
        public Result OnShutdown(UIControlledApplication application)
        {
         return Result.Succeeded;
       }
       public Result OnStartup(UIControlledApplication application)

     {
      Searching for coded .dll and the path of graph in computer
       Establish Program Panel
     }
     private void AddPushButton(RibbonPanel panel)
     {
       Establish Code for Button
     }
     private void AddSplitButton(RibbonPanel panel)
     {
       Establish Code for Pull Down Menu
     }
   }
}
Code for selectional evacuation mode:
using System;
using Autodesk.Revit.UI;
using Autodesk.Revit.DB;
using System.Windows.Forms;
namespace lineLength
{
  [Autodesk.Revit.Attributes.Transaction(Autodesk.Revit.Attributes.TransactionMode.Ma
nual)]
  public class CslineLength : IExternalCommand
  {
  public Result Execute(ExternalCommandData commandData, ref string message,
ElementSet elements )
     {
foreach (Element elem in collection)
MessageBox.Show("length of escape route"+total length calculated+"meter\n"+
  "escape timing"+total timing+"sec", "Revit", MessageBoxButtons.OK);

return Result.Succeeded;
}
public Parameter FindParameter(Element element)
```

```
{
Set null
foreach (Parameter parameter in element.Parameters)
}
derived ength feedback}
 }
}
Code for designated evacuation mode:
using System;
using Autodesk.Revit.UI;
using Autodesk.Revit.DB;
using System.Windows.Forms;
//System.Windows.Forms
namespace RevitAPI1
{
  [Autodesk.Revit.Attributes.Transaction(Autodesk.Revit.Attributes.TransactionMode.Ma
nual)]
     public class Class1 : IExternalCommand
     {
        public Result Execute(ExternalCommandData commandData, ref string message,
ElementSet elementset)
        {  Acquire length of designated id
           Parameter setup for total length acquired
           MessageBox.Show("escaping      time"+esult      derived+"sec",      "Revit",
MessageBoxButtons.OK);
           return Result.Succeeded;
        }
     }
  }
```

4 Results

The established panel are as shown in Fig. 6.

The heat release rate from analysis results is as shown in Fig. 7, the temperature derived from analysis results is as shown in Fig. 8, and the designated CO results is as shown in Fig. 9.

With the hand book of SFPE [9] of fire protection engineering, personnel safety thresholds as shown in Table 1 were set to assess the results from API calculated time for evacuation in fire simulation. Table 2 shows the results of evacuation time-derived API plug-in which were calculated using formula.

5 Conclusions

This study has utilized the Revit modeling of BIM software. The main purpose is to familiarize the secondary development of Revit, i.e., Revit API. The writing of

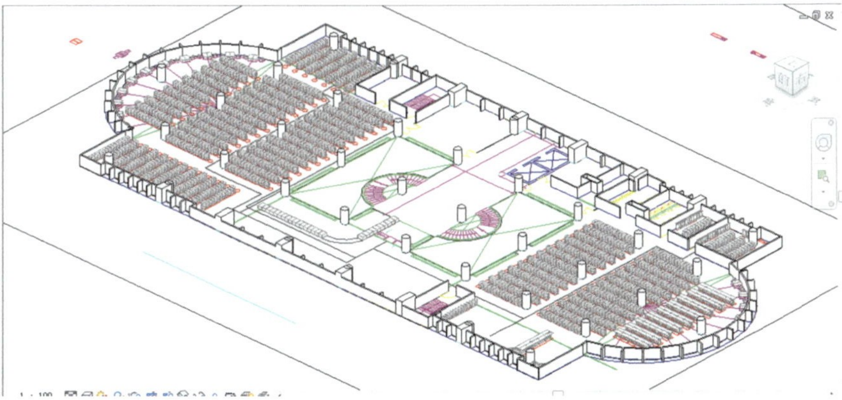

Fig. 6 Escape plug-in panel established

Fig. 7 The heat release rate derived from simulation

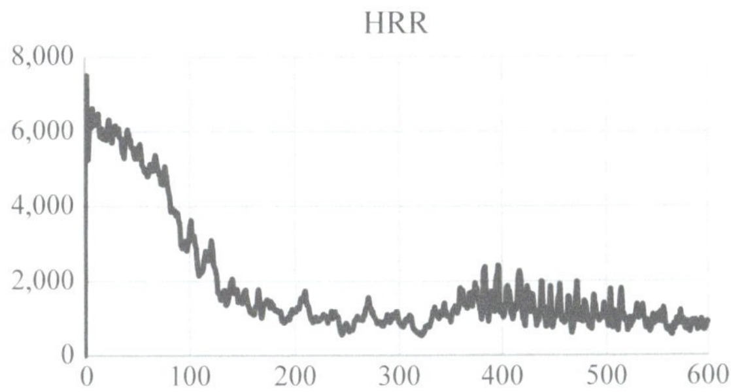

Fig. 8 The temperature from simulation

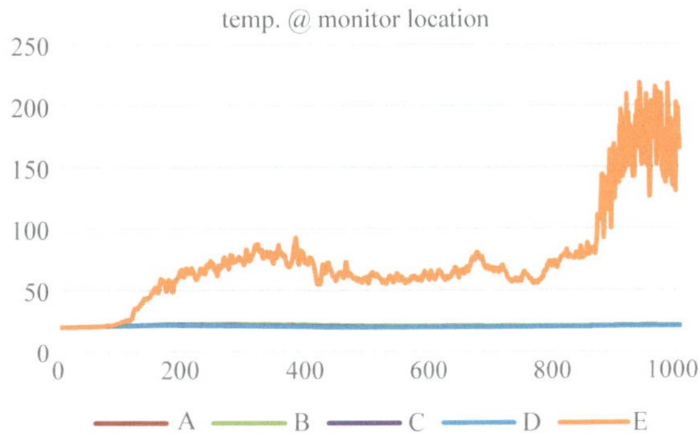

the Revit API is able to capture the data attributes of the components within Revit through designated or manual selection method, and to carry out the calculation of the presented results. In this study, the disaster evacuation aspect of the study was combined with the Revit API. The most matured personnel evacuation software and disaster prevention software in the industry today such as: Comparison of SImulex

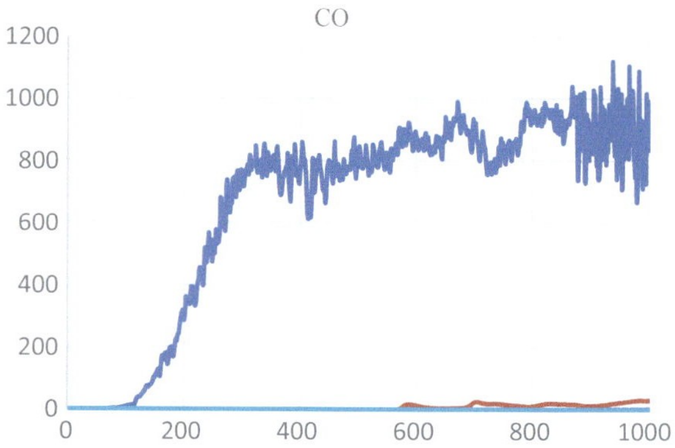

Fig. 9 The CO content derived from simulation

Table 1 Endurance and safety limits from SFPE hand book

Chem. prod.	5 min		30 min	
	Limit	Death	Limit	Death
CO (ppm)	6000–8000	12000–16000	1400–1700	2500–4000
CO_2 (%)	7–8	>10	6–7	>9
O_2 (%)	10–13	<5	<12	<7

Table 2 Results of evacuation time derived API plug-in

Fire scenes	Evacuation time (minutes)		Simulated allowable time (minutes)		
	100 persons	300 persons	CO	CO_2	O_2
Fire scene 1	5.2	12.8	19.3	17.3	26.7
Fire scene 2	5.5	12.9	21.1	18.8	34.8
Fire scene 3	5.3	12.8	19.1	18.0	28.3

from IES with FDS have been compared. All comparisons are needed to be introduced into the modeling and the format conversion within will also generate certain result deviations. In comparison, the use of Revit API can reduce the deviation from modeling conversion and is more convenient as well as relatively accurate with the results.

The purpose of this study is to utilize the Revit API to capture the data within the Revit modeling. Only the personnel evacuation route is captured and the calculation is carried out. Many objective factors such as multiple characteristics, gender, and age have yet to be considered. Hope that the following recommendations can be developed in the future:

1. The calculation formula can include factors on gender, characteristics, and age.
2. Familiarize further with the Revit API, just select only one point from the Revit modeling, and detect the edges of the walls through a method similar to as proposed, to calculate the fastest and nearest escape time.

3. Utilize mobile app to carry out three-dimensional observation and analyze the escape results of personnel affected by disasters facilitated with the display of routes on convenient spots on the mobile phone.

References

1. Wang SH, Wang WC, Wang KC, Su SY (2015) Applying building information modeling to support fire safety management. Autom Constr 59:158–167
2. Wang KC, Shin SY, Chan WS, Wang WC, Wang SH, Gansonre AA, Liu JJ, Lee MT, Cheng YY, Yeh MF (2014) Application of building information modeling in designing fire evacuation—a case study. In: Proceedings of the 31st international symposium on automation and robotics in construction and mining, Sydney, Australia, pp 9–11
3. Lin M-Y (2012) Expand building information modeling to disaster prevention and rescue applications: use of Revit platform
4. Jia T-B, Su G-W, Wei J-M, Li C-F, Chen Y-X, Lu M-Q (2012) Impact of indoor display and exit characteristics toward personnel evacuation. Hungkuang Period (66):138–158
5. Chen J-X, Yang Z-Q. Fire performance-based design for large arenas and discussion of the Taipei Dome as an example
6. Zhuang H-X (2012) Building data modeling of fire disasters and application of root cause investigation process
7. Autodesk Developer Network. http://usa.autodesk.com/adsk/servlet/index?siteID=123112&id=2484975
8. Togawa K (1955) Study of fire escapes basing on the observation of multitude currents, Report no. 14, Building Research Institute, Ministry of Construction, Tokyo
9. Society of fire protection engineers, handbook of fire protection engineering

Compressive Strength of Peat Soil Treated with Waste Tyre Granules

Nor Hazwani Md Zain and Muhammad I'zaz Zulastry

Abstract Peat is a highly organic and very compressible material, which is unable to support any building structure. It is formed in a waterlogged condition, which has limited oxygen supply. Owing to population growth, construction on this land can no longer be avoided for building developments. The construction industry is now looking forward to sustainable solutions. Therefore, the usage of waste tyres is seen as a potential replacement additive that can reinforce peat compared to other ground treatments such as cement or coarse aggregates. Nowadays, the disposal of waste tyre rubber has become a major threat to the environment globally. This paper investigates the compaction and unconfined compression strength characteristics of peat when treated with 5, 10, and 15% waste tyre granules. The results were then compared with untreated peat sample. It is observed that the maximum dry density increases and the optimum moisture content decreases when waste tyre was added at increasing amount. The stiffness of peat also increases with increasing tyre dosage as higher undrained shear strength is obtained. From this study, it is possible to use waste tyre granules as a fill material to improve the compaction and strength properties of peat soils.

Keywords Peat · Tyre · Compaction · Organic soil · Waste

1 Introduction

Peat is a dark brown and black fragmented organic soil formed in a waterlogged environment. It is derived from incomplete decomposition and disintegration of sedges, trees, mosses, and other plants growing in wet places and marshes, which is limited in oxygen supply [1]. The water content can reach up to 2000% depending on the location of peat, while organic content is normally more than 75%. Classification of peat is normally described based on the degree of decomposition and fibre content. Degree of decomposition involves a visual examination, which describes the nature of the material extruded upon squeezing. Meanwhile, the amount of fibre content

N. H. Md Zain (✉) · M. I. Zulastry
Faculty of Civil Engineering, Universiti Teknologi MARA, 40450 Shah Alam, Selangor, Malaysia
e-mail: hazwani_zain@yahoo.co.uk

M. Awang and M. R. Meor M Fared (eds.), *ICACE 2019*,
Lecture Notes in Civil Engineering 59, https://doi.org/10.1007/978-981-15-1193-6_21

in peat is described based on the amount of plant fragments that are retained on a 150 μm sieve size with a distinct cellular structure [2].

This soil is very problematic in nature due to its high compressibility and low in shear strength, which makes it difficult to support building load. Therefore, engineers tend to avoid doing construction works on this soil. Excavating and replacing or displacing the peat soil is one way of resolving the problem. Apart from that, chemical additives, compaction control, stone columns, lightweight fills and surface mattresses, surcharge loadings, pre-loading, and thermal methods have also been adopted when dealing with this difficult soil [1, 3]. However, some of the methods used have been criticised for being costly, ineffective, and unsustainable to the environment. For example, remove and replace, compaction control, stone column, and surcharge loadings are rather expensive and additives used in the stabilisation method such as lime produces calcination process, which emits carbon dioxide emission and requires high energy consumption [4], while methods involving cement production increasingly produce harmful carbon dioxide, which causes global warming. In addition, some of these methods may also require a favourable pH environment, which can be difficult to control in a highly organic soil such as peat due to its acidic condition.

Nowadays, the increasing quantities of waste materials, shortage of landfill spaces, and lack of natural earth materials have now become very alarming. Hence, there is an urgency in finding alternative ways to recycle and reuse waste materials instead of being totally dependent on natural resources, which have become scarce. It has been reported that 250 million tyres are discarded annually (approximate one tyre per person) and the latest figures are not showing any sign of declining [5]. In Malaysia, it is estimated that 8.2 million or 57,391 tonnes of waste tyre are being produced annually [6]. Waste tyre granules can be used as a substitute for aggregates in engineering fill. It has high strength (especially when steel belted), low density, high durability, and the material is in high supply at a low cost. In geotechnical engineering, waste tyres have been used in different applications, such as retaining structures, earth embankments, asphalt pavement and paving system, foundation beds, and others [7].

Studies have shown that waste tyre is able to improve the strength behaviour of soil. The authors of [8] studied the effects of including different amounts of shredded tyre chips (5, 10, 15, and 20%) in samples of peat soil with a fixed amount of sand. The results indicated that peat treated with 10% shredded tyre chips and sand had the highest unconfined compressive strength. Compaction tests were also performed by [6] and the results indicated that peat tyre-treated samples showed a reduction in maximum dry density (MDD) and an increase in optimum moisture content (OMC) as the amount of shredded tyre increases. Studies by [9] showed that at a high percentage of crumb rubber and using larger crumb rubber size, higher maximum dry density with no significant variation in optimum moisture content was observed in treated silt of high plasticity. An increasing trend of OMC and MDD is observed in clay of high plasticity soil when larger crumb rubber was used.

Peat is a complex material that requires detailed investigation as its behaviour changes depend on the degree of decomposition. Limited studies have been reported to understand the peat behaviour using waste materials such as tyre. Therefore, the aim of this paper is to investigate the inclusion of waste tyre as a replacement material in peat in terms of compaction and strength behaviour.

2 Properties of the Materials Used

2.1 Peat Soil

Peat soil was collected from a site located at Jalan Johan Setia, Klang, Selangor. The site location is shown in Fig. 1. The sample is a brownish black-coloured peat with a moderate degree of decomposition and recognisable plant structure. The water content is within the range of 200–700%, which is typical for peat in Peninsular Malaysia [10]. The organic content exceeds 75%, which refers to peat [2]. From the fibre content, the peat is considered as hemic peat for having fibre content around 33–67% [11]. The pH of peat is considered very acidic according to [2]. The physical properties of the material are summarised in Table 1.

Fig. 1 Location of peat sampling at Jalan Johan Setia, Klang (blue box)

Table 1 The physical properties of peat sample with relevant testing standards

Physical and chemical properties	Results	Standard
Soil classification	Peat	ASTM-D4427-07
Moisture content	606.8%	BS:1377:2 (1990)
Specific gravity	1.57	BS:1377:2 (1990)
Organic content	87.09%	ASTM-D2974-07a (Method C)
Fibre content	50.12%	ASTM-D-1997-91
Degree of humification	H5-H6	ASTM D5715-00
pH	4.08	BS:1377:3 (1990).

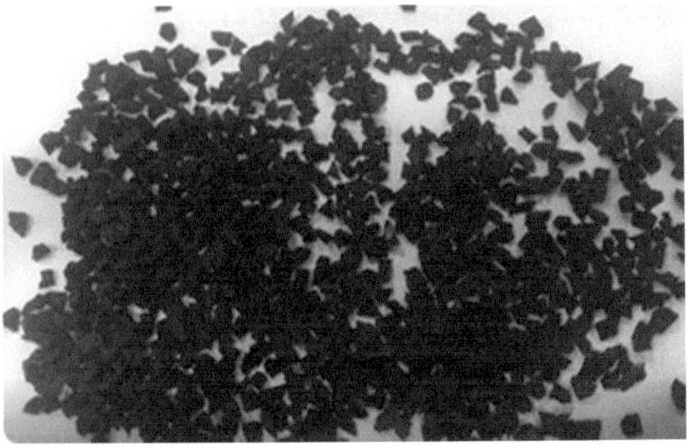

Fig. 2 Recycled waste tyre granules obtained from grinded scrap tyres

2.2 *Waste Tyre*

Recycled waste tyre granules containing 0.03% steel wires and nylon fibres passing 5–8 mm sieve with a density of 1.15–1.20 g/cm^3 were used as reinforcing material. The tyre granules were collected from a reclaimed rubber industry located in Selangor. Figure 2 shows the sample of waste tyre granules used in this research.

3 Sample Preparations and Test Methodologies

3.1 *Preparation of Compaction Sample*

Four groups of samples were prepared; natural peat and peat mixed with 5, 10, and 15% tyre granules. Natural peat soil was first air-dried on a tray until reaching at least 50% moisture content. Then, air-dried peat was sieved on a 2 mm mesh sieve aperture in order to remove large particles. Drying organic material in a high-temperature oven should be avoided in order to prevent degradation of the organic matter [12]. Particles passing 2 mm sieve size were then used for preparing the admixtures. The consistency of the sampling is very important for peat as the moisture content can easily dry out during handling. Hence, the water content of the sample before mixing should always be checked regularly in order to prevent any data inconsistencies. Once the sample was air-dried to the point of the required water content, around 1.5 kg of peat soil was thoroughly mixed with 5, 10, and 15% waste tyre granules using a hand glove for 10 min.

3.2 Standard Proctor Compaction

The standard Proctor compaction test was carried out in accordance to BS1377-4-1990 standard in order to determine the optimum moisture content (OMC) and maximum dry density (MDD) for all the samples. Each sample was compacted using 2.5 kg rammer with a tamping height of 300 mm in three layers, where each layer was subjected to 27 blows. For each compaction sample, water was added at an increment of 5% from the weight of soil. For each stage of water increment, some mixed samples were taken for water content determination.

3.3 Preparation of Unconfined Compressive Strength Sample

All samples were compacted at their optimum moisture content in the compaction mould with extension collar and base plate. Each sample was compacted in three layers where for each layer, 27 blows were applied by using a rammer weight of 2.5 kg. Compacted soil was then extruded according to the size of unconfined compressive strength where the sample dimensions were 100 mm in height and 50 mm in diameter as specified in BS1377-4-1990.

3.4 Unconfined Compressive Strength Test

The unconfined compressive strength test was performed according to BS1377-7-1990. The test was conducted at zero cell pressure. For each sample, the maximum compressive stress was determined in order to determine the undrained cohesion of untreated and treated peat samples. Figure 3 shows the sequence of testing starting from the stage of preparation of the sample.

(a) (b) (c)

Fig. 3 **a** Peat is mixed with waste tyre granules. **b** Compacted treated peat with compaction mould is held on the extruder. **c** UCS sample is extracted from the compacted tyre treated peat sample and then transferred to UCS testing machine

Table 2 The optimum moisture content and maximum dry density for each tested sample

Type of sample	OMC (%)	MDD (mg/m^3)
Natural peat	125	0.494
Natural peat + 5% waste tyre granules	123	0.52
Natural peat + 10% waste tyre granules	118	0.542
Natural peat + 15% waste tyre granules	100	0.56

4 Experimental Results and Discussion

4.1 Compaction Test

The results of compaction test on natural peat and tyre treated peat samples are shown in Table 2. It seems that the optimum moisture content (OMC) decreases when the amount of tyre granules added in the sample increases. However, the maximum dry density (MDD) increases with an increasing amount of tyre granules. Waste tyre granules replace the air voids in peat which results in a higher density after compaction. However, the optimum moisture content (OMC) reduces when the percentage of tyre in soil increases. This shows that waste tyre has low absorption capacity.

4.2 Unconfined Compressive Strength Test

Figure 4 shows the relationship between unconfined compressive strength and axial strain for untreated peat and tyre-treated peat samples. The unconfined compressive strength of each sample was first obtained by determining the maximum compression stress that the soil can sustain before it fails. Then, the undrained cohesion was obtained by plotting Mohr circles at zero cell pressure. The results in Table 3 indicate that higher undrained cohesion is obtained when the percentage of waste tyre added in the soil is higher. Peat sample, which contains 15% waste tyre granules, shows the highest unconfined compressive strength and undrained cohesion as this amount shows the highest maximum dry density.

5 Conclusion

The compaction and compression properties of peat after being treated with waste tyre granules were investigated using laboratory tests. The samples were prepared by adding waste tyre granules in the amounts of 5, 10, and 15% by weight of soil. The

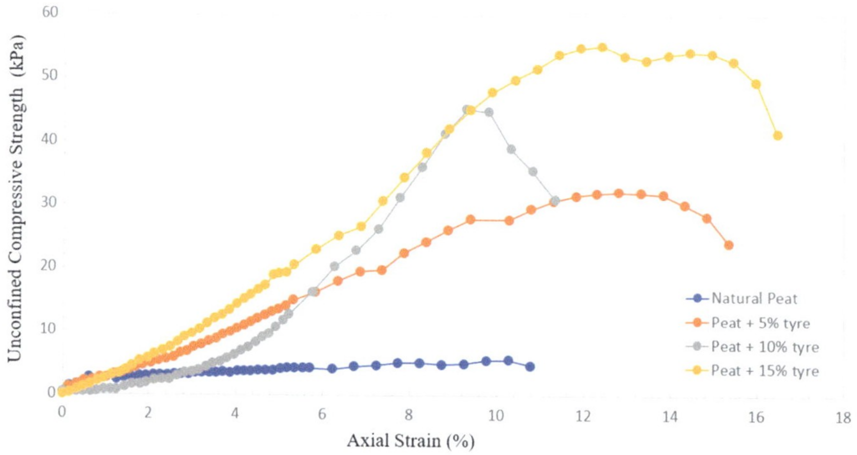

Fig. 4 Stress versus strain curves for untreated peat and treated peat samples

Table 3 The results of unconfined compressive strength and undrained cohesion for each tested sample

Type of sample	UCS value	Undrained cohesion, c_u (kPa)
Natural peat	6	3
Peat + 5% waste tyre granules	32	16
Peat + 10% waste tyre granules	45	22.5
Peat + 15% waste tyre granules	55	27.5

compaction test results show that higher percentage of tyre granules contributes to higher maximum dry density (MDD) and lower optimum moisture content (OMC). Meanwhile, the unconfined compressive strength test results show that higher percentage of waste tyre added in peat results in higher undrained cohesion. Further research is needed to investigate the optimum tyre amount that is suitable in peat samples having different organic contents and degrees of decomposition. Other geotechnical properties tests have been planned to be conducted in the future to better understand the behaviour of peat mixed with waste tyre.

References

1. Kazemian S, Huat BBK, Prasad A, Bargchi M (2011) A state of art review of peat: Geotechnical engineering perspective. Int J Phys Sci 6:1974–1981
2. Huat BBK, Prasad A, Asadi A, Kazemian S (2014) Geotechnics of organic soils and peat. CRC Press
3. Rahgozar MA, Saberian M (2016) Geotechnical properties of peat soil stabilised with shredded waste tyre chips. In: Mires Peat, pp 1–12
4. Jawad IJ, Taha MR, Majeed ZH, Tanveer AK (2014) Soil stabilization using lime: advantages, disadvantages and proposing a potential alternative. Res J Appl Sci Eng Technol 8:510–520
5. Lepcha KH, Agnihotri AK, Priyadarshee A, Yadav M (2014) Application of tire chips in reinforcement of soil: a review. J Civ Eng Environ Technol 1:51–53
6. Nik Daud NN, Yusoff ZM, Muhammed AS (2015) Ground improvement of problematic soft soils using shredded waste tyre. Jordanian Int Civ Eng Conf 6:1–5
7. Ratnam AV, Prasad DSV, Prasada Raju GVR, Sowmya Ratna P (2016) Influence of waste tyre rubber chips on strength and settlements of soils. Int J Eng Innov Res 5:2277–5668
8. Saberian M, Rahgozar MA (2016) Geotechnical properties of peat soil stabilised with shredded waste tyre chips in combination with gypsum, lime or cement. Mires Peat 18:1–16
9. Ajmera B, Tiwari B, Koirala J (2016) Geotechnical properties of clays modified with recycled crumb rubber. Geotech Struct Eng, Congr
10. Zain NHM, Mustapha M, Abdul Rahman AS (2019) Settlement behaviour of peat reinforced with recycled tyre granules. In: International Conference Built Environment Engineering—Enhancing Construction Industries Through IR4.0, pp 1–5
11. Huat BBK, Kazemian S, Prasad A, Barghchi M (2011) State of an art review of peat: general perspective. Int J Phys Sci 6:1988–1996
12. BS 1377-2, BS 1377: Part 2: 1990 (Soils for civil engineering purposes: Classification test), Br. Stand. Inst. 3 (1990)

Safety Misbehaviour and Its Effect Towards Safety Performance of Construction Projects

I. Othman and A. Azman

Abstract The construction industry, mainly the infrastructure projects, in the world reported the highest frequency of industrial accidents, mostly among the frontline workers. Domestic and international disaster-related reports indicate that the workers have high risk associated with poor safety behaviour as a major cause of accidents. Those accidents resulted in inferior project performance with works delays and financial losses due to safety procedures that need to be followed, such as project stop-work orders and payment for injury treatment cost for the workers involved. Therefore, this study assesses the causal factors of safety misconduct in construction projects that might affect the safety performance through the application of structured questionnaire and case study of an infrastructure construction project. The statistical techniques including relative importance index (RII) and average index (AVI) were used to analyse the data gathered, while the Statistical Package for the Social Sciences (SPSS) was used to measure the Spearman's rank correlation between four different groups of respondents, the Cronbach's alpha (reliability test), and validity of the study. Based on the highest impact factors identified, a framework of safety behaviour control management was proposed to improve the safety performance at the construction project. The study concludes with the conclusion on the data attained and recommendations for the improvement of safety practices in construction projects.

Keywords Construction · Safety behaviour · Causal factors · Risk

1 Introduction

Construction is a complex activity where various stakeholders are present, working under constant challenge by demands of the job. Each job will have several safety and risk factors, requiring quality and safety management systems to be established as indicated by Mehta and Agnew [1]. In Malaysia, the construction industry has

---

I. Othman (✉) · A. Azman
Universiti Teknologi PETRONAS, 32610 Seri Iskandar, Perak, Malaysia
e-mail: idris_othman@utp.edu.my

M. Awang and M. R. Meor M Fared (eds.), *ICACE 2019*,
Lecture Notes in Civil Engineering 59, https://doi.org/10.1007/978-981-15-1193-6_22

advanced since independence from British on 31 August 1957. Nowadays, the construction sector is a very important and productive sector of the Malaysian economy. Since 1970s, Malaysia, a fast-developing nation in Asian region, has realized the vital role of the construction sector, which not only contributed to economic growth but also augmented the quality of life and living standards of citizen. Construction industry is also an important industry that plays a vital role in the socio-economic growth of a country. Economically, it contributes to significant improvement in the overall gross domestic product (GDP) of a country. The industry has been consistently contributing approximately 3–5% towards the national GDP [2].

There is an increment in growth of construction, which is from 6 to 15% since the 1970s until middle 1990s [3]. Nevertheless, despite these contributions, the construction industry has faced with a wide range of challenges, which are the frequent occurrences of accidents at the workplace. To be compared with other industries, the construction industry is five times riskier [4]. The safety of workers is a complex phenomenon since construction is always dicey because of outdoor operations, work at heights, complicated on-site plants, and equipment operation coupled with workers attitudes and behaviours towards safety. Several risk factors including workers' behaviour have been suggested to have influence on the general safety at the workplace. Therefore, bad workers' safety behaviour can be considered as one of the major factors which affect the overall project performance as it causes time and cost overruns. Hence, a suitable safety management plan should be done thoroughly to reduce the safety misbehaviour by the workers, which has a significant impact on project performance.

Despite the fact that improvements in construction safety have been made over the last decade, accidents and injuries still occur on construction sites from time to time and it appears that construction safety has reached a plateau. These accidents and injuries have resulted in huge personal, social, and economical costs, and also resulted in an unsatisfactory project performance of works' delay and financial losses due to safety procedure that needs to be followed such as project stop-work orders by safety officer and payment for injury treatment costs for the workers involved.

To prevent accidents and poor project performance, considerable attention has been paid by the author to explore their root causes. A classic work of Heinrich's Domino Theory (1931) understood accidents as linear outcomes of unsafe conditions and human errors. It was claimed that over 88% of preventable accidents were caused by unsafe behaviours [5]. Such an understanding has led to a traditional view on human error, which is a cause of accidents [6]. Hence, the unsafe behaviour or safety misconduct constitutes the most significant factor towards low safety performance of any construction project.

This project focuses on some of the objectives, which are:

- To identify and rank the causal factors of workers' safety misbehaviour during the construction stage which contributes largely to the safety performance and triumph of overall project based on relative importance index (RII) with correlation between construction practitioners.

- To validate the factors and construction practices influencing the safety performance during the construction phase through the application of project case study.

The scopes of study in this project are as follows:

This research is to identify, rank, and analyse the causal factors of workers' safety misbehaviour based on the actual case study of an infrastructure construction project. The application of gap analysis system in this research tends to improve the validity of it through identifying the relative factors that may have not been explored yet by past researches.

The construction industry has constituted the highest rates of casualties than other industries worldwide [7–12]. HSE states that the construction industry brings significant contribution and benefit towards facilities production, initiates various economic activities, and enhances the social and environmental needs of a nation [13]. Thus, construction safety becomes one of the significant concerns.

Safety performance was traditionally measured according to the record of accident or injury data [14]. Hallowell et al. [15] stated that actions were being taken in response to adverse trends in injuries. Measures regarding the rate of accidents and compensation costs can either be forthcoming or after the event and are relatively uncommon [16]. Focusing on safety results, the success of safety is indicated by levels of system failure [14]. However, most of the safety professionals and researchers agreed that delaying indicators about an accident at workplace or injuries of people must occur before a measure can be done. It may or may not provide the necessary awareness to prevent future accidents [17].

The construction industry continues to account for a disproportionate number of occupational fatal and nonfatal injuries [18]. However, in the last decade, regardless of the continuous safety efforts, the injury rates by construction sector have increased. According to the Bureau of Labor Statistics (2012), the US construction sector has experienced more than 1000 fatal injuries every year between 1995 and 2008 [19]. In 2011, construction workers accounted for a fatality rate of 9.1 per 100,000 full-time equivalent (FTE) workers, as opposed to the all-worker fatality rate of 3.5 per 100,000 FTE workers [20]. These injury statistics clearly prove that construction workers are more likely to lose their lives or to be injured as compared to workers from other industries.

According to National Safety Council (NSC; as cited in Mitropoulos et al. 2005), "an unsafe condition or activity that, if left uncontrolled, can contribute to an accident" is known as hazard [21]. To prevent injuries, hazard recognition methods are introduced at which procedural or physical controls are being used; thus, workplace hazards can be identified and risk associated with these hazards can be minimized. Hazards that can be detected during assessment may have less control in place; consequently, it may lead to severe threat to both safety of workers and the environment. Thus, it is very crucial to execute an organized effort to identify and evaluate processes and activities for potential hazards. Those kinds of both informal and formal methods provide important information. From the information obtained, the safety problems can be addressed and operational risks can be managed.

In construction industry, improper safety behaviour is known as one of the serious problems. The number of accidents in construction project is increasing proportionally with a vast development of this industry, which then, will affect the safety performance. Therefore, it is very vital to identify the factors and construction practitioners' practices that influence the safety behaviour of the workers in construction projects in order to overcome the problems stated, and prevent fatalities either directly or indirectly. The main identified factors are: (1) worker involvement factor, (2) material and equipment factor, (3) workplace factor, and lastly, and (4) management involvement factor.

2 Methods

Descriptive method, which involved collecting data in order to answer the questions concerning the subject of the research, was chosen as a method for this study. This is because as mentioned by Fraenkel and Wallen [22], descriptive method is a method used to explain, analyse, and classify something through various techniques, survey, interview, questionnaires, observation, and text. This type of research is used to find the answers to who, what, where, and how questions. It was designed to portray the characteristics of a population (Saunders et al. 2007).

There are four methodology phases for this research; literature review, case studies, empirical studies, and evaluation. The purpose of this study is to identify the causal factors of workers' safety misbehaviour and rank them. While mainly looking into construction industry in Malaysia, the researcher is trying to understand better about the industry by describing the current performance of the Malaysian construction sector through study. Those will help the researcher to identify the areas of problems and try to provide corrective measures towards the safety problems. Thus, this study can be explained as descriptive.

The next step is the distribution of questionnaire survey among construction professionals including engineers, safety and health officers, site supervisor, and others which have been involved in the construction project. According to Hill [23] by increasing the sample size, more data can be collected, and the statistical power is improved. Based on Roscoe's Simple Rules Of Thumb, Roscoe (1975) in most research recommended samples of 30 or more, whereas the use of statistical analyses with samples less than 10 is not recommended [24]. Therefore, there are 30 respondents who have been participated throughout the survey.

3 Results and Discussion

A total of 35 questionnaires were administered for this survey, of which 30 were returned with valid responses. The questionnaires were distributed through online form method. This showed a response rate of 86%. From the results attained, it was observed that majority of the respondents believed that they were aware of the importance of good safety behaviour in a construction project in order to have high safety performance and prevent occupational fatalities and injuries.

3.1 Lack of Supervision

Lack of supervision was ranked as the first highest impact factor with highest AVI and RII. Lack of supervision for material and equipment-related factor might be due to lack of safety awareness among the contractor practitioners. This problem then led to the unsafe attitude and behaviour towards the material and equipment handling. There is no proper guidance to the workers regarding their work. Therefore, it is needed for the worker to be briefed and the supervisor should be responsible in monitoring the worker. This is because not all of the workers have holistic knowledge regarding safety control practices and they tend to do their work with the wrong sequence and method. Based on the problem, it is advisable to the contractor company to enhance the safety knowledge of the workers by giving internal trainings, regular briefings, and also by providing the safety booklets that can be understood easily by the workers.

3.2 Lack of Worksite Inspection

The second highest impact factor for the implementation of safety control on resources is the lack of worksite inspection, which is categorized under workplace-related factor. Lack of worksite inspection might occur due to irresponsible contractor practitioners; therefore, safety awareness should be developed among them, and more importantly, to the top management. Lack of worksite inspection also might be due to lack of budget provided to hire fulltime safety personnel. However, worksite inspection is very crucial to provide safety controls since the construction workers are exposed to hazards more than other industries. The workers are involved in concreting work, plastering work, and others which are very risky due to the changing conditions and risky areas. Since the worksite inspection was rarely been done, the workers tend to disobey the rules in construction project. For example, they tend to not wear proper PPEs and perform their task in an unsafe way. Besides, the critical surroundings such as the void areas will not be properly covered or secured and the scaffolding will not be inspected. Therefore, it is needed to have site assessments and react properly to the wrong actions, so the potential hazards can be identified and can be addressed before any unwanted incidents occur.

3.3 Unsafe Behaviour and Attitudes

With the third highest AVI (4.000), unsafe behaviour and attitude have been categorized under the high impact factor. This problem occurrence might be due to lack of knowledge among the workers. The workers tend to act based on their own way, which is opposed to the correct method. Besides, this problem also happens because the workers receive minimum guidance for safety control practices. Therefore, it is really important for the worker contractor to provide safety education programmes and trainings for the worker. Besides, for the implementation of the safety programme, full commitment from top management is required.

3.4 Poor Housekeeping

Housekeeping is one of the significant ways to keep the site's environment in the satisfying state. To improve the housekeeping, the site should be inspected by safety and health officers. Besides, the site should be cleaned to have a proper access to the construction site and to smoothen the construction workplace. Poor housekeeping will lead to various unsatisfying condition, such as the stagnant water in the walkway area leading to breeding ground for mosquitoes and can cause electrocution if an electrical appliance was exposed to the water. Besides, the poor housekeeping also causes the construction site to be exposed to the abundance of wastes in the site, causing site constraint to workers. This can lead to serious accidents such as fall from height. Therefore, regular inspection should be done and supervisors should monitor the site condition.

3.5 Lack of Knowledge

Another top five highest impact factors affecting the implementation of safety control on resources in construction project is lack of knowledge. This lack of knowledge leads to various problems. Without adequate knowledge, the worker will exhibit unsafe behaviour upon their work and they would also have poor housekeeping. The contractor would have low safety awareness towards the hazards. However, it is very vital for the worker to be provided with proper education and training regarding the safety. The worker should be monitored by providing the adequate input to enhance their daily work routine. Thus, reduces the risk at the workplace.

4 Conclusions

This study has identified the top high impact causal factors of workers' safety misbehaviour practices in construction projects. "Site supervisor" and "safety and health officers" have rated "worker involvement related factors" as the highest impact factor compared to other groups of factors, while the engineers have mostly chosen the "material and equipment related factors" following with "worker involvement related factors" and "workplace related factors" as the highest impact factors. According to the contractors' practitioners, the least important factor is "management involvement-related factors". Based on the analysis, the top five highest impact causal factors of workers' safety misbehaviour practices are: (i) lack of supervision, (ii) lack of work-site inspection, (iii) unsafe act and attitudes, (iv) poor housekeeping, and (v) lack of knowledge. The framework on safety behaviour control management is developed. The highest impact causal factors of workers' safety misbehaviour practices in construction project can be determined and several control measures are suggested in order to overcome the highest rank root causes.

References

1. Mehta RK, Agnew MJ (2010) Analysis of individual and occupational risk factors on task performance and biomechanical demands for a simulated drilling task. Int J Ind Ergonom 40(5): 584–591
2. Malaysia C (2009) Construction industry review 1980–2009 (Q1). Construction Industry Development Board (CIDB), Kuala Lumpur
3. Raftery J, Pasadilla B, Chiang Y, Hui EC, Tang B.-S (1998) Globalization and construction industry development: implications of recent developments in the construction sector in Asia. Constr Manag Econ 16(6):729–737
4. Sawacha E, Naoum S, Fong D (1999, October) Factors affecting safety performance on construction sites. Int J Proj Manag 17(5):309–315
5. Heinrich H, Superintendent A (1930) Relation of accident statistics to industrial accident prevention. Proc Casuallity Act Soc 16(33–34):170–174
6. Dekker S (2004) Ten questions about human error: A new view of human factors and system safety. CRC Press
7. Bansal V (2011) Application of geographic information systems in construction safety Planning. Int J Proj Manag 29(1):66–77
8. Statisitcs BOL (2015) Census of fatal occupational injuries (CFOI) - Current and revised data. https://www.bls.gov/iif/oshcfoi1.htm#charts. Accessed 2 July 2015
9. Statisitcs BOL (2015) Census of fatal occupational injuries (CFOI) - Current and revised data 2013 chart package. https://www.bls.gov/iif/oshwc/cfoi/cfch0012.pdf. Accessed 13 July 2015
10. Brunette MJ (2004) Construction safety research in the United States: targeting the Hispanic workforce. Inj Prev 10(4): 244–248
11. HKOSA Health (2015) Occupational safety and health statistics. https://www.labour.gov.hk/eng/osh/pdf/archive/statistics/OSH_Statistics_2012.pdf. Accessed 2 July 2015
12. HAS Executive (2015) Health and safety in construction in Great Britain. https://www.hse.gov.uk/statistics/industry/construction/construction.pdf. Accessed 2 July 2015

13. HAS Executive (2015) Phase 1 report: underlying causes of construction fatal accidents—a comprehensive review of recent work to consolidate and summarise existing knowledge. https://www.hse.gov.uk/construction/resources/phase1.pdf. Accessed 2 July 2015
14. Choudhry RM, Fang D, Mohamed S (2007) Developing a model of construction safety culture. J Manag Eng 23(4):207–212
15. Hallowell MR, Hinze JW, Baud KC, Wehle A (2013) Proactive construction safety control: Measuring, monitoring, and responding to safety leading indicators. J Constr Eng Manag 139(10):04013010
16. Cooper MD, Phillips RA (2004) Exploratory analysis of the safety climate and safety behavior relationship. J Safety Res 35(5):497–512
17. Grabowski M, Ayyalasomayajula P, Merrick J, Harrald JR, Roberts K (2007) Leading indicators of safety in virtual organizations. Safety Sci 45(10):1013–1043
18. Findley M, Smith S, Kress T, Petty G, Kim E (2004) Safety program elements in construction. Prof Saf 49(2):14
19. Short KS, Garner TI, Bureau U (2012) The supplemental poverty measure a joint project between the census bureau and the bureau of labor statistics by
20. Pegula S, Janocha J (2013) Death on the job: fatal work injuries in 2011
21. Mitropoulos P, Abdelhamid TS, Howell GA (2005) Systems model of construction accident causation. J Constr Eng Manag 131(7):816–825
22. Fraenkel JR, Wallen NE, Hyun HH (2011) How to design and evaluate research in education. New York: McGraw-Hill Humanities/Social Sciences/Languages
23. Hill R (1998) What sample size is "enough" in internet survey research. Interpers Comp Technol: an Electr J 21st century 6(3–4):1–12
24. Roscoe JT (1975) Fundamental research statistics for the behavioral sciences [by] John T. Roscoe

Causes of Government Physical Project Delays in Prime Minister's Department, Malaysia

Nor Aida Idrus, Syuhaida Ismail and Abd Latif Saleh

Abstract Most of the government physical projects are not completed on time as specified in the contract due to the award of extension of time (EOT). As of the current Eleventh Malaysia Plan (11 MP), the numbers of delayed projects keep increasing and the delayed factors have remained the same. Most of these delays can be avoided if the stakeholders at all stages of project implementation pay more attention during the planning and implementation stages of the government physical project. Therefore, this paper aims to appraise the causes of government physical projects delays in Prime Minister's Department, Malaysia. Via a questionnaire survey on 100 officers from Prime Minister's Department as the project's client, Public Work Department (PWD) as the project's consultant, and Implementation Coordination Unit (ICU) as monitoring agency, it is found that the causes of delay in government physical projects are due to consultant-related delay factors, client-related delay factors, and contractor-related delay factors, consisting of poor consultant's communication with other stakeholders, poor consultant's coordination with other stakeholders, and poor consultant's site management, as well as client's financial difficulties, client's changes in design, and client's tardiness in making decisions; and also contractor's financial difficulties, problem of subcontractor, and shortage of contractor's labour, respectively. The findings of this paper are beneficial towards proposing the area of improvement that can be translated into specific strategies towards on-time government physical projects delivery in Malaysia.

Keywords Extension of time · Physical project · Construction · Prime Minister's Department · Public Work Department

N. A. Idrus (✉) · S. Ismail
Razak Faculty of Technology and Informatics, Universiti Teknologi Malaysia Kuala Lumpur, Wilayah Persekutuan Kuala Lumpur, 54100 Kuala Lumpur, Malaysia
e-mail: noraidaidrus@gmail.com

S. Ismail
e-mail: syuhaida.kl@utm.my

A. L. Saleh
School of Civil Engineering, Faculty of Engineering, Universiti Teknologi Malaysia, Johor Bahru, 81310 Johor, Malaysia
e-mail: alatif@utm.my

M. Awang and M. R. Meor M Fared (eds.), *ICACE 2019*,
Lecture Notes in Civil Engineering 59, https://doi.org/10.1007/978-981-15-1193-6_23

1 Introduction

The government of Malaysia remains as the largest client in the construction industry, focusing mainly on the building of public facilities and infrastructures, such as schools, hospitals, clinics, public housings, and roads. The construction industry requires substantial financing [8]. For the overall implementation under the Eleventh Malaysia Plan (11 MP), the Malaysian government allocates a development expenditure ceiling of RM260 billion for the period between 2016 and 2020 [3]. The total for the government's commitment to spearhead National Development from First Malaysia Plan (1 MP) to Tenth Malaysia Plan (10 MP) is RM875.2 billion [3].

Despite the allocation by the government, construction delays are still apparent, especially during the implementation stage of the government physical projects. Delay in delivering government physical projects on schedule has become a serious problem for the stakeholders involved. Based on Development Project Performance Report by Implementation Coordination Unit (ICU) [6] of Prime Minister's Department, Malaysia, most of the government physical projects were not completed on time as specified in the contract management and were further awarded the extension of time (EOT). ICU [7] also reported that 61 physical projects were delayed due to problems encountered at implementation stage and 87% are delayed at construction stage. Since physical project delays are seen as a common phenomenon for government physical projects, even though government has implemented almost 13 national development plans, this paper aims to appraise the causes of government physical projects delays in Prime Minister's Department, Malaysia, towards on-time government physical projects delivery in Malaysia.

2 Literature Review

Delay in government physical projects can be defined as delays exceeding one month or 10% later than the expected schedule [5]. Hamzah et al. [4] defined delay as time overrun or extension of time to complete the project. To place more emphasis, delay in construction physical project can also be expressed as the time overrun or extension of time behind the date agreed upon by the contract parties [11]. There are 25 causes for construction physical projects delays, which are further grouped into consultant-related delay factors, client-related delay factors, and contractor-related delay factors, as shown in Table 1.

On the other hand, the causes of delays for five government physical projects under Prime Minister's Department have been also determined, as illustrated in Table 2, namely: (1) improper planning and scheduling, (2) contractors and consultants are not competent in implementing projects, (3) lack of effective communication and coordination between parties, (4) estimated actual cost exceeds the allocated budget, (5) changes in design, and (6) slowness in approving major changes in scope of work [10].

Table 1 Causes of construction physical projects delays by key participants of physical construction projects

Contractor-related delay factors	Client-related delay factors	Consultant-related delay factors
1. Shortage of labours 2. Labour productivity 3. Materials delivery on site 4. Resources shortage (material, machinery, equipment) 5. Financial difficulties 6. Inadequate experience 7. Subcontractor problems 8. Construction mistakes and defective work	1. Financial difficulties 2. Slowness in progress payment 3. Changes by owner during construction 4. Failure to provide the required construction site 5. Slowness in making decision by the owner	1. Professional team productivity 2. Shortage in professional team 3. Increment of material's price 4. Slowness in approving major changes in scope of work 5. Problems during inspection 6. Poor site management 7. Improper planning and scheduling 8. Changes in design/design error 9. Slow preparation and approval of shop drawings 10. Incomplete documents 11. Lack of effective communication and coordination 12. Building permit approval

Source Yusof et al. [13], Othman and Ismail [9], Alias [1], Srdić and Šelih [12], Zidane and Andersen [14], Hamzah et al. [4], Samarah and Bekr [11]

Table 2 Causes of delays for physical construction projects under Prime Minister's Department

Project	Delay and extension of time (EOT)	Causes of delays
A	530 days delay from the original completion date Numbers of EOT: 5	1. Improper planning and scheduling 2. Contractors and consultants are not competent in implementing projects
B	EOT 1 approved for 59 days	1. Lack of effective communication and coordination between parties
C	219 days delay from the original completion date EOT 1 approved for 219 days	1. Estimated actual cost exceeds the allocated cost 2. Lack of effective communication and coordination between parties
D	16% delayed from actual schedule	1. Changes in design
E	120 days delay from the original completion date EOT 1 approved for 120 days	1. Changes in design 2. Slowness in approving major changes in scope of work

3 Research Methodology

This paper used quantitative approach via document and literature review as well as questionnaire survey. Extensive literature review on the data of government's physical projects under Prime Minister's Department from government reports and related studies is conducted in recognising the exact research area. As for the questionnaire survey, the questionnaire went through a pilot run to 10% of expected sample [2] to test the reliability and validity of the questions before being used for the actual data collection. The questionnaire is distributed to three groups of stakeholders in government physical projects encompassing 100 respondents, namely Prime Minister's Department as the project client, Public Work Department (PWD) as the project consultant, and Implementation Coordination Unit (ICU) as the monitoring agency. Finally, expert focus group interview is conducted on five ICU officers that have experience and knowledge in project management more than 10 years to validate the data found and analysed from the questionnaire survey.

4 Results and Discussion

The results of analysis on the questionnaire on the causes of government physical project delays are shown in Table 3, where in general the most important delay factor is consultant, followed by client, and finally contractor-related delay factor based on the mean value.

Based on the results for RII analysis in Table 4 for the consultant-related delay factors, it is found that only 16 out of 17 causes are practically significant and valid, where the top 5 most important consultant-related delay factors are poor communication with other stakeholders, poor coordination with other stakeholders, poor site management, improper scheduling, and improper planning. As for the client-related delay factors, there are six causes identified based on importance ranking consisting of financial difficulties, changes in design, slowness in making decision, improper scheduling, failure to provide required construction site, and slowness in progress payment, while for the contractor-related delay factors, all proposed causes are deemed important by respondents and the top 5 most important contractor-related delay factors are financial difficulties, problem of subcontractor, shortage of labour, poor site management, and inadequate experience.

Table 3 Results of mean value analysis

	Delay factors of government physical projects in Malaysia	Minimum	Maximum	M ± SD
1	Consultant-related delay factor	1.00	5.00	3.74 ± 0.92
2	Contractor-related delay factor	1.00	5.00	3.55 ± 0.85
3	Client-related delay factor	1.00	5.00	3.65 ± 0.97

Table 4 Results of RII analysis

Priority	Factors	RII (%)
Consultant-related delay factors		
1	Poor communication with other stakeholders	81.0
2	Poor coordination with other stakeholders	81.0
3	Poor site management	80.6
4	Improper scheduling	80.0
5	Improper planning	79.6
6	Poor project management practices	77.6
7	Changes in designs	77.2
8	Changes in specifications	77.0
9	Slowness in approving major changes in scope of work	74.4
10	Slow approval of shop drawings	72.0
11	Consultant team non-productivity	71.8
12	Inaccurate estimated project cost	71.8
13	Inadequate experience of consultant	71.4
14	Incomplete documents	71.2
15	Slow preparation of shop drawings	70.8
16	Shortage in consultant team	64.2
Client-related delay factor		
1	Financial difficulties	75.0
2	Changes in design	75.0
3	Slowness in making decisions	74.8
4	Improper scheduling	74.4
5	Failure to provide the required construction site	69.8
6	Slowness in progress payment	69.4
Contractor-related delay factor		
1	Financial difficulties	88.2
2	Problem of subcontractor	80.4
3	Shortage of labours	79.8
4	Poor site management	79.4
5	Inadequate experience	78.8
6	Poor coordination with other stakeholder	78.6
7	Poor communication with other stakeholder	78.4
8	Poor project management practices	78.0
9	Delay of delivery materials on site	75.8
10	Labour non-productivity	73.6
11	Construction defective works	73.4
12	Shortage of materials	70.6
13	Construction mistakes	69.0

5 Conclusions

Based on the statistical analysis and data validation via expert focus group interview, it is concluded that three factors of delay, namely consultant-related, client-related, and contractor-related delay factors are the main factors which contribute to government physical projects delays in Malaysia. It is also agreed that the consultant and client are ranked first and second, respectively, in the delay factor table as they are involved in all three phases of the project cycle, which are planning, implementation, and post-implementation, whereas the contractor, on the other hand, comes in later which starts from the implementation phase. These delay factors can be used as the guidelines in coming up with the improvement measures or specific strategies towards on-time government physical projects delivery.

Acknowledgements The authors would like to express their sincere gratitude to the Ministry of Higher Education Malaysia, Universiti Teknologi Malaysia (UTM), and the Research Management Centre (RMC) of UTM for providing the financial support for this paper to be published. This study is financed by the Research University Grant Tier 1 under the Cost Center No. Q.J130000.2522.19H54.

References

1. Alias I (2006) Faktor-faktor kelewatan dalam projek pembinaan bangunan pendidikan MARA. Masters thesis, Universiti Teknologi Malaysia, Faculty of Civil Engineering
2. Connelly LM (2008) Pilot studies. Medsurg Nurs 17(6):411–412
3. Economic Planning Unit (EPU) (2016) 11th Malaysia plan strategy paper
4. Hamzah N, Khoiry MA, Arshad I, Badaruzzaman WHW, Tawil NM (2012) Identification of the causes of construction delay in Malaysia. In: Proceedings of world academy of science, engineering and technology, no 72, p 614)
5. Icu JPM (2015) Pengurusan Projek Awam: Konsep, Praktikal dan Realiti. Unit Penyelarasan Pelaksanaan Jabatan Perdana Menteri, Putrajaya
6. ICU JPM (2017) Project development performance report of RP3- RMK11
7. ICU JPM (2018) Project development performance report of RP3-RMK11
8. Ministry of Finance (MOF) (2013) Economic report
9. Othman A, Ismail S (2014) Delay in Government project delivery in Kedah, Malaysia. Recent advances in civil engineering and mechanics 248–254
10. Prime Minister's Department (2019). http://www.jpm.gov.my/jpm2/
11. Samarah A, Bekr GA (2016) Causes and effects of delay in public construction projects in Jordan. Am J Eng Res (AJER) 87–94
12. Srdić A, Šelih J (2015) Delays in construction projects: causes and mitigation. Org Technol Manag Constr: Int J 7(3):1383–1389
13. Yusof MA, binti Mohammad N, bin Mat Derus Z (2007) Excusable and compensable delays in the construction of building project–a study in the states of Selangor and Wilayah Persekutuan Kuala Lumpur, Malaysia. Change 49(40):71
14. Zidane YJT, Andersen B (2018) The top 10 universal delay factors in construction projects. Int J Manag Proj Bus 11(3):650–672

Improving Thermal Comfort Through Natural Ventilation and Passive Solar Systems in Residential Buildings in Iraq: Review Paper

Mustafa Abdulmunem Saleh Al Mohsen, Sumarni Ismail and Ida Suriana Ismail

Abstract The building enterprise in Iraq performs the greatest instance of energy, especially in residential section. Indeed, most of the residential constructions rely on traditional construction systems. In the past, the traditional system used to supply thermal comfort for residents during summer and winter seasons. However, due to the fact of the growing impact of global warming and technological advancement, this constructional system is no longer viable. Therefore, realizing energy efficiency, specifically in Iraq, is very vital due to non-stop electricity crisis since the 1990s. Hence, this review paper is an attempt to show several strategies to achieve thermal comfort in residential buildings in Iraq through natural ventilation and passive solar systems for both cooling and heating periods, which at the end can help to achieve some kind of sustainability in residential buildings in Iraq, specifically, hot climates in general.

Keywords Natural ventilation · Energy efficiency · Sustainability · Hot dry climates · Thermal comfort

1 Introduction

Iraq is characterized by very high temperatures in the summer and moderate in winter. The highest values in June, July, and August are between 43 and 50 °C and in January between 1 and 8 °C [1]. Winds and water cause natural hazards such as dust storms. There are two types of winds and dust storms in Iraq. From June to September, the country is affected by dry air masses from the Mediterranean [2]. This leads people to use air-conditioners excessively during the summer to bring suitable thermal comfort environment inside their desired space. Besides, there is a shortage in electricity in Iraq [3], which causes the power to get shutdowns as a result of power consumptions especially by air-conditioners. Hence, roof is one of the most important elements of a

M. A. S. Al Mohsen (✉) · S. Ismail (✉) · I. S. Ismail
Faculty of Design and Architecture, UPM, Seri Kembangan, Malaysia
e-mail: mostafaalrubaiee@gmail.com

S. Ismail
e-mail: sumarni@upm.edu.my

M. Awang and M. R. Meor M Fared (eds.), *ICACE 2019*,
Lecture Notes in Civil Engineering 59, https://doi.org/10.1007/978-981-15-1193-6_24

building. Roof, walls, and ventilation are the most important parameters in a building that contribute to heating load in old and non-insulated buildings [4]. Therefore, if we control thermal gaining through ceilings, we can provide thermally comfortable indoor climate, reduce air-conditioner usage and power consumption, reduce cooling demands, and improve current houses' energy efficiency to achieve sustainable status. Thus, this study is a review on improving thermal comfort through ventilation and passive solar systems inside residential buildings by considering Iraq as a part of hot and dry climatic zone.

2 Thermal Comfort Problems in Hot and Dry Climates

The fundamental purpose of a building is to shield the residents from the outside circumstances, such as heat in summer, cold in winter as well as wind and rain. Hence, an applicable indoor climate should be designed for buildings. An international issue on source of power and environmental problems is prompted to reflect on consideration of many factors for the design of buildings. According to a large number of research works, in order to indicate the climate and its suitability for the building design and building materials, designers and/or architects have to be aware of the characteristics of the local weather in their working environment. Furthermore, based on the related climatic elements, they will be capable of categorizing the construction problems and observe or endorse their solutions to keep away from them.

In order to design a building with thermal comfort and suitable indoor climate in hot climates, primarily, the common construction problems have to be recognized. The following elements are commonly referenced in many studies:

- High temperature
- High solar radiation
- Moisture or high RH level
- Excessive heat gain in summer
- Heat loss during winter.

Therefore, to acquire the stated goal, architects as construction designers, have to take a look at some of the factors in the design stage, namely construction layout, orientation, and envelope details (shape, insulation, solar control) to manage overall thermal performance and sustainability of buildings [5–13].

3 Construction Issues in Overheated Periods

In summer, in regions with hot climates, construction must be designed in such a way to decrease the amount of heat gain. To reflect on consideration on this issue, distinctive aspects of the construction should be analysed by considering two important factors: solar control and thermal insulation. For instance, for the purpose of controlling the solar radiation, considering a suitable construction layout, orientation, and shape will be really useful and will make it possible to take advantage of strategies such as shading, envelope reflective texture, and transparent envelope. In the meantime, making use of thermal insulation will grant a kind of barrier to isolate the indoors from outdoors, and consequently, it can manipulate the consumption of energy and cost. As mentioned above, solar management is one of the major techniques in construction design in hot and dry climates. In order to have a successful manipulation on solar radiation, the following parameters have to be taken into account precisely [8–10]:

- Building layout, form, and orientation
- Building envelope
- Opaque envelope (including walls and roofs)
- Transparent envelope (including walls and roofs).

4 Passive Solar Systems

Different techniques of designing and constructing a passive system can be classified into two groups with two essential strategies: managed heat gain strategy or passive heat gain, and passive cooling method. Each of these techniques can be divided into four simple systems: direct gain, thermal storage wall, attached sunspace, and convective loop (Fig. 1).

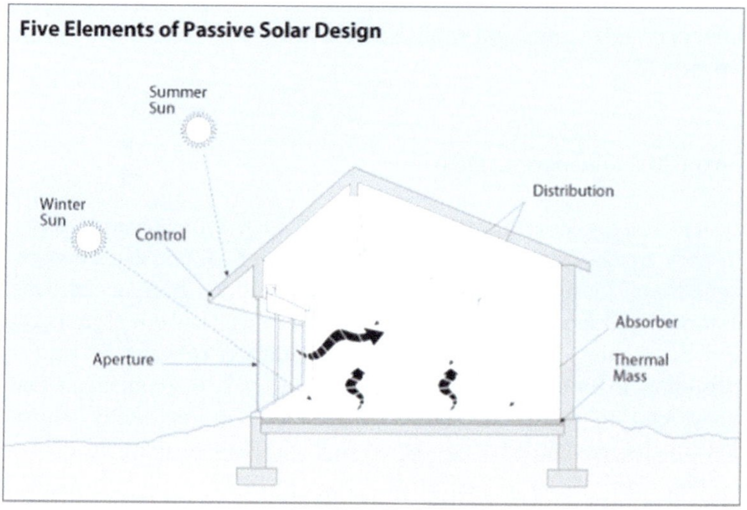

Fig. 1 Five elements of passive solar system (*Source* URL evereco.org/sustainable_building.html)

Each one of these fundamental passive systems consists of five collectively dependent fundamental components: collector, absorber, storage, distributor, and manipulator [8]. Here, each strategy and its associated subsystems will be described.

4.1 Passive Heating Systems

As noted above, we have four distinct passive methods. In this part, each method will be reviewed to evaluate its functionality in heating seasons briefly.

In direct gain systems, in cold season, sunlight radiates into construction through the collector, which is pointed towards south and typically made of glass. This solar radiation will be transformed into heat via elements, which have absorbed it. The heat is then either used to warm the region or stored to be used later. Normally, it is difficult to provide direct solar radiation over the whole surface area. So reflecting sunlight from light-coloured surfaces to the dark surfaces will be an acceptable solution [8].

4.1.1 Convective Loop Systems

In this system, sun rays will be transferred via the collector of the convective loop of the thermo-siphon air panel (TAP) and strike the absorber surface. The absorber has a metallic surface with back colour, which converts solar radiation to heat. It should be located in two subsystems, which are vertical and U-tube panels. In vertical panel, in order to avoid warm room air from being drawn back into the panel, simple backdraft dampers are generally supplied at the vents. But, in U-tube panels, shut proximity of the inlet and outlet vents will furnish a benefit for this system in comparison to vertical panels (Fig. 2) [8].

4.1.2 Attached Sunspace System

In this system, sun rays pass through the collector, which is directed towards the south and then, they are absorbed by elements and is transformed to heat. This procedure is the same for all its subsystems such as open wall, direct gain, air exchange, and thermal storage wall subsystems. For instance, in open wall and direct gain subsystem because of no opening, there is a direct and free transfer of warm air between two spaces. In order to avoid immoderate loss of heat, it is recommended that high-performance glazing be used, or in direct achieve systems, the use of transportable insulation which is positioned at the shared wall will limit the extra heat losses from building (Fig. 2) [8].

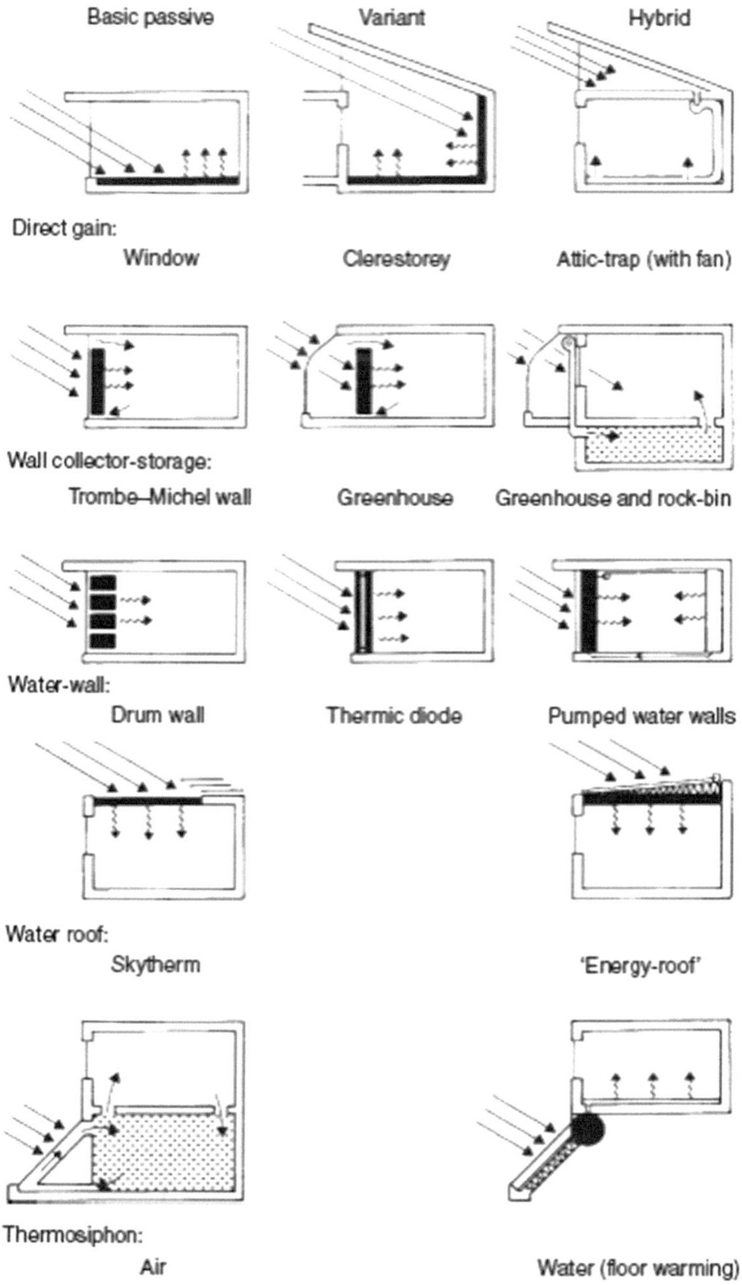

Fig. 2 Main types of passive solar heating systems [19]

4.1.3 Thermal Storage Wall System

In summer, solar radiation comes in via the collector, hits the storage wall, and makes it warm. In the state of unvented thermal storage wall, the generated heat is saved and slowly transferred to the interior. Because of the tendency of hot air to go up and cold air to come down, in the vented wall system, a natural ventilation system will supply hot air to the building. Although using vents will provide light and some direct heat gain to the building during the day, however, they will also decrease the efficient zone of the storage wall (Fig. 2) [8].

4.2 Passive Cooling (Ventilation) Systems

As noted at the start of this chapter, the fundamental feature of the construction is the supply of barrier from outdoor temperature circumstances to evade thermal comfort problems. One of the factors which commonly create thermal comfort problems is relative humidity stage and moisture content of the construction envelope. Since in hot and humid climate normally there is an extra quantity of steam in the air, therefore, in cold season, this quantity of steam will be absorbed through construction envelope and at some point in hot season, it will be launched again to the air from construction envelope. Thus, extra quantity of moisture content in construction envelope could be an enough resource for distinct kinds of issues in the case of thermal comfort or structural defects. These sorts of defects could be listed as condensation, structural warmth loss, and moisture motion through affecting sturdiness of construction materials, which will be mentioned later [14–19].

On the other hand, immoderate quantity of steam in the air will dramatically impact the thermal comfort stage in buildings. Since the pressure of the moisture in the air relies solely on the humidity or moisture content, therefore, the steam pressure of the extra humid air inside a hot building will be greater than the steam pressure of even saturated outdoor cool air. Increasing the level of humidity could extend sweating and make life difficult. Therefore, ventilation turns into an essential function to avoid immoderate humidity. In this case, it should be provided to allow moisture in the warm internal air diffuse to the exterior through the influence of the difference in vapour pressure. There are various approaches to supply air ventilation. Using of passive or energetic systems should grant them. In summer, via stratifying some controls to passive techniques, an advantage of natural ventilation and passive cooling systems could be delivered to the building. Those controls can be described as follows.

4.2.1 Attached Sunspace System

In hot periods, this system can simply be overheated. Thus, the collector should be properly shaded to benefit from transportable insulation. In addition, sun space should be vented properly. It is extremely suggested to set low vents in the shades on the windward side of the system, and top vents on the leeward side. In addition, the storage mass in this system can also assist to cool construction if precisely designed and controlled (Fig. 2) [8].

4.2.2 Convective Loop Systems

Since convective loop systems are only intended for warming, they will be inactive throughout the hot season (Fig. 2) [8].

4.2.3 Thermal Storage Wall System

In summer, collector has to be properly shaded and resized to prevent immoderate heat gain. In case of wall vents, vents are positioned on the exterior side of the collector element in shape of opaque panels. Moreover, transportable insulation should be used to shield the collector from outer side and preserve the cool temperature of the wall as much as possible (Fig. 2) [8].

4.2.4 Direct Gain System

Through the overheated periods, collectors should be properly shaded to keep away from excessive warmth gain. Shading elements, such as overhangs, should be sized precisely to increase general overall system performance (Fig. 2) [8].

5 Constructing Issues in Under-Heated Periods

Passive solar systems are based on the precise design of the organization of the building's areas and suitable choice of building materials in order to acquire heating and cooling advantages from the natural free power resources to minimize electrical energy wastes via air conditioning systems [8].

In areas with hot climates, the use of passive solar systems will add the gain of solar heat to the building and decrease the power consumption. Because of their significance in under-heated periods, they have a dominant function in order to provide natural ventilation in overheated periods.

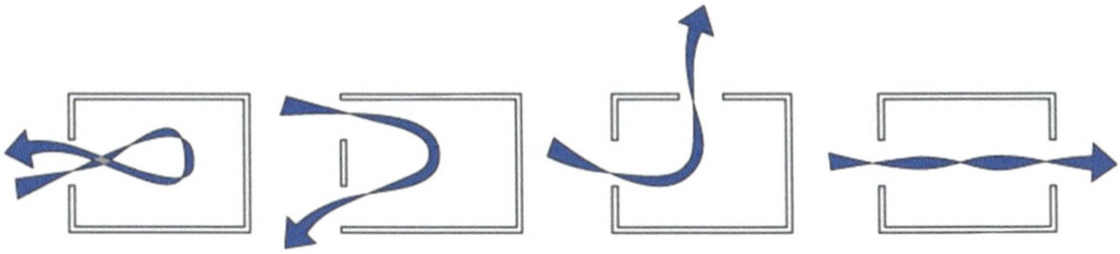

Fig. 3 The ventilation and openings positions in a plan [18]

Table 1 Wind velocity upon openings position [18]

Opening height as a fraction of wall height	1/3		
Opening width as a fraction of wall width	1/3	2/3	3/3
Single opening	12–14%	13–17%	16–23%
Two openings in same wall	–	22%	23%
Two openings in adjacent walls	37–45%	37–45%	40–51%
Two openings in opposite walls	35–42%	37–51%	47–65%
"Range = wind 45° perpendicular to opening"			

6 Improving Ventilation

Ventilation is optimized by providing cross-ventilation through many spaces in the house as a practical solution. In regular wind conditions, the aspect of effective to negative pressure is encouraged. Of course, the indoors design of the residence needs to allow the air to flow through and indoors doors in the ventilation route need to stay opened [20].

Ventilation effectiveness relies on wind velocity, the angularity at which the wind strikes the window, and the place and dimension of the window (Fig. 3). A room with a single opening will have solely 12–23% of the wind velocity. This improves up to 51% if openings are positioned on adjoining partitions and about 65% of the outdoor air speed can be reached with openings on contrary walls (Table 1) [20].

7 Conclusions and Recommendations in Planning, Design, and Construction Materials for Buildings in Hot Dry Climatic Zones

7.1 Conclusions

In hot and dry climatic zones, it is very important to use the available green energies such as wind and solar radiation to achieve thermal comfort for residents. Therefore, systems such as passive solar systems can provide a suitable solution for Iraq climate which is well known by the very extremist weather during summer and winter seasons. In addition, windows dimensions and positions can affect the wind velocity and the amount of humidity inside the room or the zone, which can lead to improve climate, temperature, oxygen, CO_2, and steam amount inside a room and provide better ventilation and thermal comfort for residents.

7.2 Recommendations

7.2.1 Architectural Design Recommendations

1. Using the idea of a building with an internal courtyard as a temperature regulator. Use of water and greenery in private and public yards.
2. Exploitation of flat roof surfaces as open areas (roof garden).
3. Reduce solar radiation, increase shadows, and increase humidity.
4. Use of "*shanshul/mashrabiya*" as a cooling system for buildings and also increasing the amount of shadows.
5. Use the idea of the indirect entrance to avoid climatic factors from dusty winds.
6. Avoid directing the buildings to the direction of the wind and directing the openings towards the north-west and south-west.

7.2.2 Urban Planning Recommendations

1. Use a built-in general layout that reduces the lengths of lanes and solar glare.
2. Use the style of agglomerated buildings and provide interior courtyards.
3. Use of narrow, zigzag pedestrian streets which receive the least amount of solar radiation.
4. Taking into consideration the natural terrain of the site and using plants and special trees which bear the climate conditions of hot and dry regions.
5. Pay attention to urban formations by setting heights and scales in hot and dry deserted areas.

7.2.3 Building Materials Recommendations

1. It is preferable to use materials with high thermal capacity.
2. Using prominent bricks to double the shadows on the facades.
3. Using heat insulation materials on roof tiles and between wall materials.
4. It is preferable to use a double roof to allow continuity of airflow through the building block.

References

1. Muir J. Iraq marshes face grave new threat. BBC News, 24 February 2009. http://news.bbc.co.uk/2/hi/middle_east/7906512.stm
2. Thompson Wiley (2002) Country Map, "Location". In: Malinowski Jon (ed) Iraq: a regional geography. U.S. Military Academy Press, West Point, NY, p 2002
3. Istepanian H (2014) Iraq's electricity crisis
4. Huang J, Hanford J, Yang F (1999) Residential heating and cooling loads component analysis. Lawrence Berkeley National Laboratory University of California Berkeley, CA, p 94720
5. Ikrom Zakaria NZ, Woods P (2002) Building designs for hot and humid climate. Malaya and Multimedia University, Malaysia
6. Watson D, Kenneth L (1983) Climatic building design—energy efficient building principles and practice. McGrow Hill, NewYork, USA
7. Chenvidyakarn T (2007) Passive design for thermal comfort in hot humid climates. J Archit/Plan Res Stud 5(1)
8. Crosbie MJ (1998) Steven Winter Association, The passive solar design and construction handbook. Wiley, New York
9. ROAF S (1992) Energy-efficient building: a design guide. Oxford, Blackwell Scientific Publications, London
10. Manioglu G, Yilmaz Z (2006) Energy efficient design strategies in the hot dry area of Turkey, Elsevier
11. Omer MA (2006) Renewable building energy systems and passive human comfort systems. Nottingham, UK, Elsevier
12. Lam JC, Wan KK, Tsang CL, Yang L (2008) Building energy efficiency in different climate. Xian University of Architecture and Technology, Shaanxi, China, Elsevier
13. Yilmaz Z (2006) Evaluation of energy efficient design strategies for different climatic zone: comparison of thermal performance of building in temperate-humid and hot-dry climate. Istanbul Technical University, Turkey, Elsevier
14. Hancer P (2005) Thermal insulation of roofs for warm climates. PhD thesis, Eastern Mediterranean University, North Cyprus
15. Harriman LG, Lstiburek JW (2009) The ASHRAE guide for buildings in hot and humid climates. ASHRAE, Atlanta, USA
16. ASHRAE (1997) Handbook of fundamentals. American Society of Heating, Refrigerating and Air-Conditioning Engineers Inc, Atlanta, GA
17. Lucas F, Adelard L, Garde F, Boyer H (2001) Study of moisture in buildings for hot humid climates. France
18. http://en.wikipedia.org/wiki/Relative_humidity
19. Terchsel HR (1994) Moisture controls in buildings. American society for testing and material, Philadelphia
20. DeKay M, Brown GZ (2001) Sun, wind, and light: architectural design strategies